TRIBOLOGY OF THIN LAYERS

TRIBOLOGY SERIES

Vol. 1 Tribology — a systems approach to the science and technology of friction, lubrication and wear (Czichos)

Vol. 2 Impact Wear of Materials (Engel)

Vol. 3 The Tribology of Natural and Artificial Joints (Dumbleton)

Vol. 4 Tribology of Thin Layers (Iliuc)

TRIBOLOGY SERIES, 4

TRIBOLOGY OF THIN LAYERS

IVAN ILIUC

Institute of Physics and Technology of Materials
Bucharest, Romania

EDITURA ACADEMIEI REPUBLICII SOCIALISTE ROMÂNIA
BUCUREŞTI, ROMÂNIA

ELSEVIER SCIENTIFIC PUBLISHING COMPANY
AMSTERDAM—OXFORD—NEW YORK 1980

Revised and updated version of "Tribologia straturilor subțiri"
Editura Academiei, București, 1974

Translated from the Romanian by Doina Olivia Popa

Published in co-edition with
Editura Academiei, Bucharest, Romania
Distribution of this book is being handled by the following publishers

for the U.S.A. and Canada
ELSEVIER/NORTH-HOLLAND, INC.
52 Vanderbilt Avenue
New York, N. Y. 10017, U.S.A.

for the East European Countries, China, Northern Korea, Cuba, Vietnam and Mongolia

EDITURA ACADEMIEI REPUBLICII SOCIALISTE ROMÂNIA
CALEA VICTORIEI 125,
R—71021— BUCHAREST, ROMANIA

for all remaining areas
ELSEVIER SCIENTIFIC PUBLISHING COMPANY
335 Jan van Galenstraat
P. O. Box 211, 1000 AE Amsterdam, The Netherlands

with 170 illustrations, 33 tables, and 204 references

Library of Congress Cataloging in Publication Data
Iliuc, Ivan.
(Tribology series; v. 4)
Revised and updated version of "Tribologia straturilor subțiri".
Bibliography: p.
Includes index.
1. Tribology. I. Title. II. Series.
TJ. 1075. I 4513 1980 621.8'9 79—27369

ISBN 0—444—99768—7 (Vol. 4)
ISBN 0—444—41677—3 (Series)

Printed in Romania

Preface

The process of friction between two solid surfaces is inevitably accompanied by the occurrence of wear and energy losses. If the surfaces are separated by a layer of dissimilar material, the friction process also takes place within the layer, and the wear of the surfaces and possibly the energy losses due to friction are reduced considerably. When the separating material is a fluid lubricant, the separating film is maintained by the viscosity of the fluid and the relative displacement of the surfaces. With low speeds or small film thickness, boundary or mixed film lubrication conditions exist and friction occurs either directly between the solid surfaces or between the films covering the surfaces. If the operating conditions are such that fluid lubrication is not possible, the use of thin solid films may be necessary for surface separation. Unlike hydrodynamic films, adsorbed or thin solid films are effective only if they adhere tenaciously to the surface. They are not dependent upon surface velocity but only on the presence of the substrate.

As thin surface films can influence the tribological properties of surfaces irrespective of the properties of the bulk material they are of considerable importance in the design of machine elements.

Thin films in the present context are those films for which the formation and properties are greatly dependent upon the supporting bulk material. Such films are different from thick deposited films, having properties dependent upon the characteristics of the deposited material. A clear delineation of the two types is, however, not possible.

Thin films may be subdivided into adsorbed, reaction, soft metallic, conversion and solid lubricant-formed films. The oxide layer present on metallic surfaces in air has a special significance. Specific operating conditions such as high vacuum or an inert atmosphere, which prevent regeneration of the oxide layer, affect frictional behaviour and necessitate measures to provide surface protection.

Adsorbed and reaction layers are useful in aiding hydrodynamic lubrication. They are formed by the interaction of lubricant additives with metallic surfaces. Such additives allow continuous regeneration of the layer. Besides reducing the coefficient of friction and wear rate, additives also improve seizure resistance. A better understanding of the seizure process and the mechanism of action of e.p. additives can aid the selection and the effectiveness of gear lubricants.

Another class of thin protective films is achieved by solid lubricants. They may be used with or without liquid lubricants, particularly under arduous conditions such as elevated temperature or hostile environment.

The properties of the various protective films are assessed, using simple test apparatus and test specimens or full scale test rigs. The effectiveness of the various test machines depends upon their specific characteristics and particularly on the contact conditions. Investigational techniques play an important role in the study of the structure and the mechanism of action of the films.

By presenting the fundamental aspects of the mechanism of action and the structure of the various types of thin films, this book emphasises their common properties and outlines various practical applications. The chapters dealing with homogeneous thin layers, the use of radioactive tracers in the study of wear and seizure mechanisms present original research results.

The book was originally published in 1974 and the present edition has been improved and updated.

The author wishes to thank all concerned with compilation of the original version and the English translation.

I. ILIUC

Contents

1 The solid surface

1.1 Surface topography

The real surfaces of solids, irrespective of the method of formation contain irregularities or deviations from the geometrical forms prescribed. Even the smoothest surfaces, such as those obtained by cleavage of some crystals, contain irregularities the height of which exceeds the interatomic distances by several times.

Any measurement of surface characteristics should take into account the ideal geometrical form by comparing it with the actual surface. The deviations of the actual surface with respect to the ideal geometrical surface can be divided into several classes. First order deviations are shape deviations; second order deviations are waves. The surface roughness includes third order deviations.

For technological applications the macroscopic roughness which corresponds to second and third order deviations is of primary interest. For problems pertaining to the physico-chemistry of surfaces, microroughness or higher order deviations must be considered.

1.1.1 Techniques for measuring surface roughness

Various instruments are available for the measurement of surface roughness. They may be classified according to the principle of measurement used, the range and accuracy of the measurement, or the method of recording the results.

Stylus instruments using electronic amplification are the most popular. This technique, solely recommended by I.S.O. is generally used for reference purposes and the investigation of surface topography.

Each technique has limitations regarding the minimum and maximum size of the asperities it can assess or the mean arithmetic value R_a determined. The essential features of various techniques are summarized in Figure 1.1. A lower limit is imposed by the sensitivity and discrimination capacity of the instrument and an upper limit by the measurement capacity. Except for the stylus method, the other techniques of measuring surface relief provide numerical data which can only be correlated empirically. Such techniques are mainly used for continuous inspection procedures in service as they function without contact with the surface.

The pneumatic method determines surface roughness by means of the pneumatic resistance between a circular transducer and the surface (Figure 1.2). For a constant rate of air flow the pressure drop is determined by the overall roughness

$$G = \frac{1}{\pi D} \int_0^{\pi D} y \, dx, \tag{1.1}$$

where $y = f(x)$ is the surface equation for the contact area between the transducer and the surface.

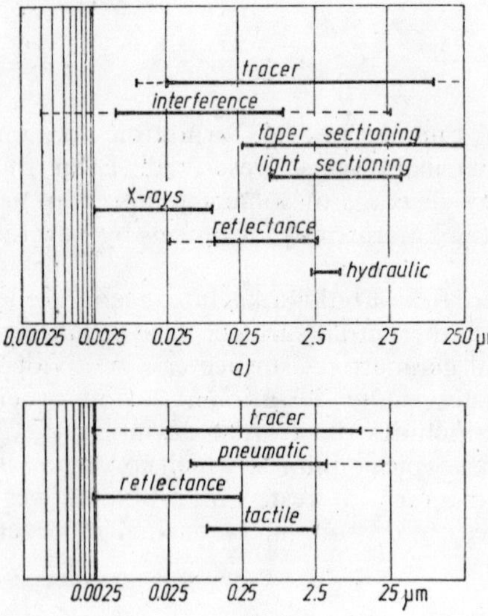

Figure 1.1 Range of utilization of roughness measurement techniques [73]: *a*, asperity height; *b*, roughness R_a.

Figure 1.2 Schematic diagram of pneumatic technique.

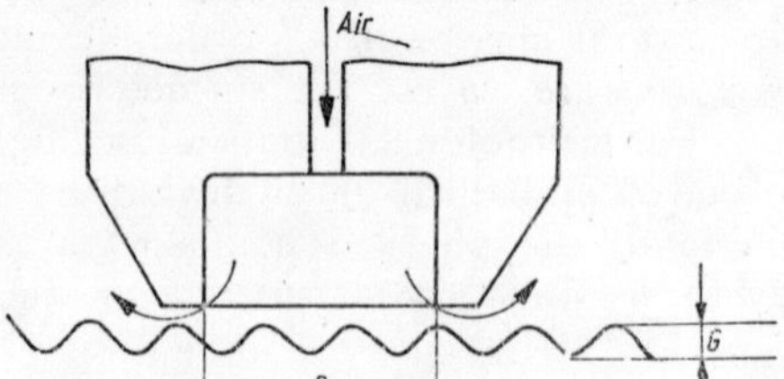

The hydraulic method measures the equivalent roughness of the surface by the time taken for a given volume of liquid to travel a given length of the surface measured. A simple relationship exists between the height of asperities and the flow time

$$R_z = at^n, \tag{1.2}$$

where a and n are constants determined by the characteristics of the method employed.

The capacity method makes use of the relationship between the capacity of a plane circular condenser and the roughness of one of its armatures. The condenser consists of a smooth surface disk and the surface to be measured.

Optical methods. Microscopical examination of a cross-section of a surface permits the estimation of the surface roughness and structural modifications in the immediate sub-surface area. To protect the cross-section during preparation for examination, the surface is usually covered with a plastic or soft metal alloy layer. The cross-section can be normal, inclined or parallel to the surface to be examined. The use of a taper section permits an extra magnification factor in a direction at right angles to the taper. This exaggerates the asperity size (Figure 1.3).

Oblique illumination of a surface for microscopical examination is a non-destructive method for enhancing the magnification in one direction (Schmaltz). In this procedure the surface is illuminated by a beam of light at an angle of incidence of 45° to the surface. The profile is examined with the aid of a microscope, also arranged at an angle of 45° to the surface.

Figure 1.3 Tapered section of roughness measurement [39].

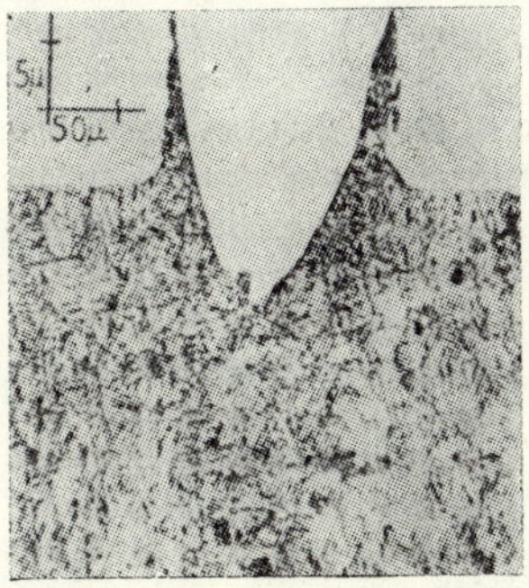

The optical reflectivity of a surface may be used to measure surface roughness. If the wavelength of the light used is much greater than the root mean square of the surface profile, then reflectance depends only on the roughness; asperity slope has no effect.

Under these conditions

$$\log \frac{\rho_0}{\rho} = \frac{(4\pi rms)^2}{2.303\,\lambda^2}, \tag{1.3}$$

where

ρ_0 is the reflectance of a standard smooth surface,
ρ — the reflectance of the surface measured,
λ — the wavelength of the light used.

Interference methods. Use of optical interference methods provides a contour map of the surface, the bands connecting points on the surface of the same height. The difference in height between the bands is half the wavelength of the light used; the geometrical shape of the surface is irrelevant.

1.1.2 **Stylus tracer instruments**

This method uses an instrument which amplifies and records the vertical motions of a stylus displaced at a constant velocity by the surface to be measured. The large amplifications required have imposed the exclusive use of electrical displacement transducers and electronic amplifiers. Two types of transducers are used:

(a) generating transducers,

(b) parametric transducers.

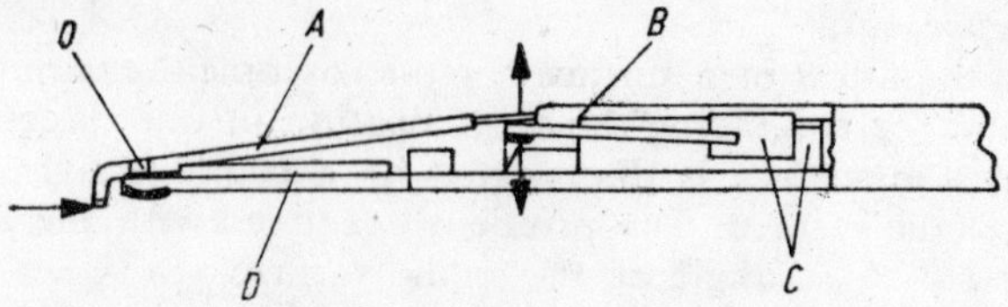

Figure 1.4 Transducer with piezoelectric crystal: *O*, hinge; *A*, level ; *B*, piezoelectric blade; *C*, couple with fluid friction; *D*, elastic blade.

The first type includes moving coil and piezoelectric transducers. The current induced in the coil located in a uniform magnetic field depends on the coil displacement speed, which is a function of amplitude and stylus displacement frequency. With a piezoelectric transducer the signal depends on the load to which the crystal is subjected. As the piezoelectric blade rigidity is high the use of a liquid friction coupling ensures the vertical displacement of the stylus (Figure 1.4).

The stylus vertical displacements are transmitted by means of the lever *A* and hinge *O* to the crystal *B*. The stylus mobility is ensured by the viscous coupling *C*. At high displacement speeds of the stylus, the coupling acts rigidly causing deformation of the crystal. The load is applied by means of the elastic blade *D*.

Parametric transducers, such as those with carrier frequency, ensure a signal determined only by the stylus vertical position to provide a signal similar to the profile measured, independent of the displacement velocity. For high accuracy recording instruments, parametric transducers are used.

To obtain a correct representation of the profile measured, the stylus support should displace on a reference surface the geometrical form of which is similar to that of the surface measured and is parallel to it. Use of an external reference surface makes the alignment of the parallelism of the two surfaces difficult. This method however provides the real surface profile (Figure 1.5 *a*). This difficulty

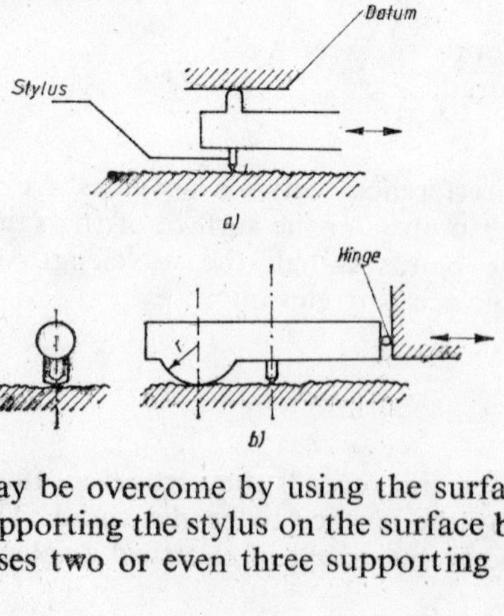

Figure 1.5 Reference surface for roughness measurement: *a*, outer surface of reference; *b*, part surface as surface of reference.

may be overcome by using the surface to be measured as a reference surface by supporting the stylus on the surface by means of a sphere (Figure 1.5 *b*). In some cases two or even three supporting points are used.

1.1.3 Sources of errors

Errors due to the stylus. To minimize wear of the stylus a spherical stylus type is used. However small the radius of curvature, some parts of the surface will not be completely traversed. For example, with a standard stylus of 2.5 μm radius the error made in the exploration of a sinusoidal roughness of total height smaller than 6 μm and pitch qreater 12μm does not exceed 10%. With this type of roughness the main distortion is due principally to the second harmonics and in the case of triangular roughness the distortion is due to the second and sixth harmonics.

Errors due to the transducer dynamic characteristics. The response frequency characteristics of the transducer indicate its ability to reproduce the surface profile faithfully. A high natural frequency is obtained by using as small a mass as possible. Moreover, the support should be maintained rigid and the friction at hinges should be as small as possible. The transducer characteristics are closely related to the force pressing the stylus against the surface. By increasing the load, accuracy can be improved but damage to the surface may occur, which may lead to erroneous results. For a stylus of 2.5 μm radius a load smaller than 1 g is generally used.

Errors due to electronic circuits. Limitation of the pass band of the amplifier leads to elimination of the influence of waviness on the roughness value. With most instruments the lower frequency cut off can be modified in steps within broad limits. Expressed as wavelength it may be varied, e.g. between 0.25 mm and 2.5 mm (Figure 1.6). The wavelength is related to frequency by means of the displacement velocity.

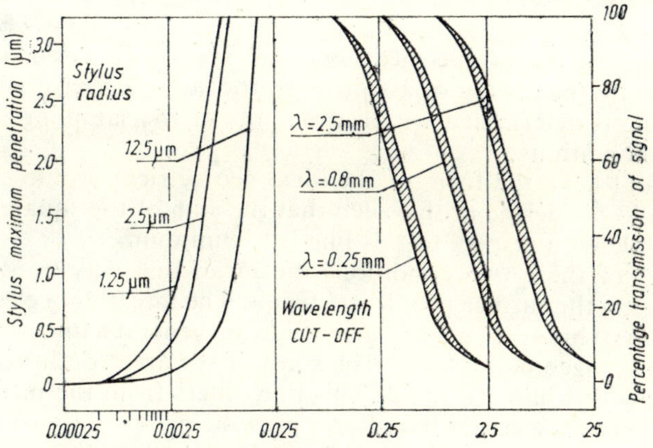

Figure 1.6 Errors of transducer-amplifier assembly for stylus roughness tester [73].

At high frequencies, that is for asperities of small width, the signal is damped by the radius of curvature of the stylus, which restricts the passage of the stylus into small depressions.

Effect of the surface datum. The use of the surface to be measured as a reference surface introduces some errors as the stylus is subjected to a combined motion. The output signal depends on the stylus itself and its support positions. The relative errors are however small and this technique is mainly employed in surface roughness measurement.

1.1.4 Estimation of the profile

Owing to the difficulties implied by the interpretation of the three-dimensional roughness and to the advantages offered by stylus measurement, the surface profile is used as a basis for further assessment of surface characterization (Figure 1.7). From the sampling length l a series of values are deduced to characterize the surface roughness. Two systems are available for evaluating the profile,

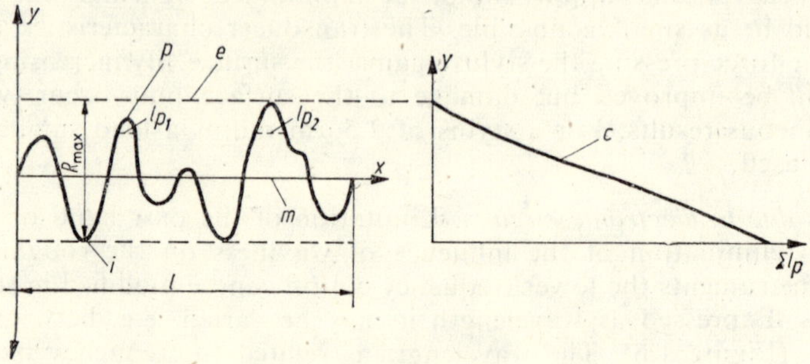

Figure 1.7 Elements of a profile in system M:
p, effective profile; l, basic length; m, mean line; e, outer line; i, inner line; c, bearing curve.

system M, based on the profile mean line and system E, based on an envelope curve generated by the rolling of a sphere on the surface measured. With system M the mean line is determined within the limits of the sampling length l by the following two conditions:

— it should have the form of the ideal geometrical profile,
— it should divide the profile such that the sum of the squares of the deviation of the profile with respect to this line is a minimum.

By means of these two conditions the mean line may be determined but plotting it by using the surface profile is difficult. The integrating electronic circuits incorporated in stylus instruments determine in general a line which is parallel to the profile macrogeometric form. The sums of the areas outlined by the profile above and below this line are equal. This line differs from the mean line for any profile and is termed the center line. For a periodic profile the two are equivalent.

By means of the mean line and the surface profile a series of values may be defined to characterize the surface.

The center line average R_a of the profile is defined as the average of the ordinates of the profile with respect to the mean line

$$R_a = \frac{1}{l}\int_0^l |y|\,dx, \tag{1.4}$$

or approximately

$$R_a = \frac{\sum_{i=1}^{i=n} |y_i|}{n} \cdot \tag{1.5}$$

As the arithmetic mean deviations may be determined with a relatively simple electronic circuit, this is the quantity generally used.

The average peak-to-valley height R_z is defined as the difference between the arithmetic mean of the ordinates of the highest five peaks and the arithmetic mean of the ordinates of the deepest five troughs of the profile. These ordinates are measured within the limits of the sampling length with respect to a line parallel to the mean line situated outside the profile

$$R_z = \frac{1}{5} \sum_{i=1}^{i=5} R_i - \frac{1}{5} \sum_{j=1}^{j=1} R_j, \tag{1.6}$$

where R_z is a value of the total depth of the surface asperities (or ten-point height) This is the parameter estimated by visual observation or microscopical measurement of the surface. It is influenced by scratches and microcracks.

The maximum height $R_{max}(R_t)$ is defined as the distance between the highest peak and the deepest valley along the sampling length.

When waviness exists, the quantities R_a and R_z are influenced by the value of the measured length l. Electrical circuits of stylus instruments allow selection of the measured length l to reduce the influence of waviness.

In addition to R_a and R_z other values may be determined in system M. These include:

— the root mean square r.m.s., defined by the relation

$$\text{r.m.s.} = \sqrt{\frac{1}{l} \int_0^l y^2 \mathrm{d}x}, \tag{1.7}$$

— the bearing length L defined by the relation

$$L = \sum l_p. \tag{1.8}$$

The bearing length, as a function of the asperity height, determines the (Abbot) bearing curve (see Figure 1.7). A knowledge of this curve permits some assessment of the wear and running-in of surfaces. From Abbot's curve it is possible to calculate the amount of material it is necessary to remove by running-in to achieve a specific bearing surface.

For an arbitrarily chosen profile there is no correlation between the various values defined above. However, it is usually considered that, R_{max}/R_a is 3 to 8. To characterize a surface, one or several values should be given.

In order to obtain a more complete characterisation of surface roughness other values have also been suggested. They can be obtained by using the first and second order derivatives of the profile (Myers [127]). It has been suggested that in addition to r.m.s. one should use the values

$$Z_2 = \sqrt{\frac{1}{l}\int_0^l \left(\frac{dy}{dx}\right)^2 dx}, \tag{1.9}$$

and

$$Z_3 = \sqrt{\frac{1}{l}\int_0^l \left(\frac{d^2y}{dx^2}\right)^2 dx}. \tag{1.10}$$

The values Z_2 and Z_3 indicate the existence of sharp edges on the surface profile. Thus, for a perfect triangular profile r.m.s. = 0.017 μm, $Z_2 = 3.73$ and $Z_3 = \infty$. If the profile peaks are rounded off, then r.m.s. = 0.015 μm, $Z_2 = 3.16$ and $Z_3 = 0.44\ \mu m^{-1}$. It may be observed that although both profiles have the same r.m.s. the values of Z_2 or Z_3 differ.

1.1.5 **Statistical characteristics of the surface profile**

Stylus instruments (of parametric type) provide an electrical signal analogous to the surface profile measured. The use of an analog-digital converter and a computer permits the complex processing of the information contained in the primary signal.

Using such a technique Williamson and Hunt [180, 79] found that the distribution of the profile heights as well as the distribution of peak heights for most modes of machining is Gaussian (Figure 1.8). Deviations from this distribution are more significant in the lower parts of the profile. This permits the use of a Gaussian distribution in all problems related to contact between surfaces, as the approach of surfaces takes place in all cases above the mean line. Another important conclusion was that the height of an asperity and its radius are not of the same order of magnitude as is often assumed. For example an asperity of height 0.1 μm may have a radius of 1 mm.

The stationary and ergodic statistical character of the surface profile permits its description by the following parameters:

(a) center line average and root mean square,

(b) the height distribution function,

(c) the auto-correlation function,

(d) the spectral density function.

The center line average and the root mean square, for a profile and surface can be readily determined. For the remaining characteristics an analog-digital converter coupled to a computer must be used.

The auto-correlation function $R_{xx}(\lambda)$ may be obtained by displacing the profiles $X(l)$ by the distance λ and averaging the product of the two profiles [194].

$$R_{xx}(\lambda) = \frac{1}{N - \lambda} \sum_{i=1}^{N-\lambda} \mathring{X}(l_i)\mathring{X}(l_i + \lambda), \tag{1.11}$$

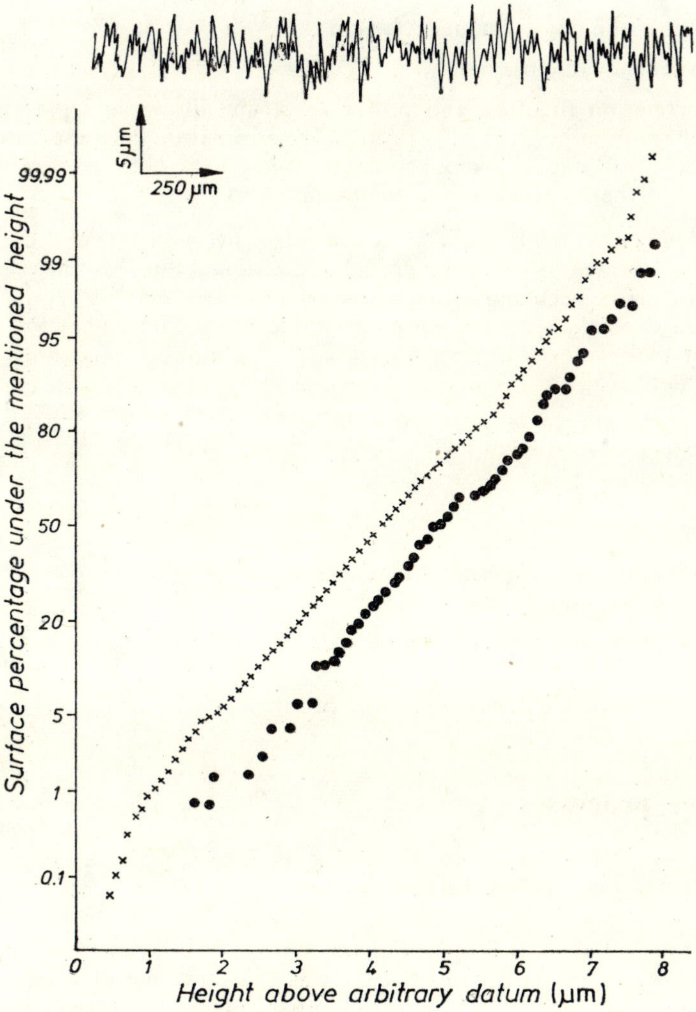

Figure 1.8 Distribution of heights (+) and peaks (·) for a steel surface [79].

where

$\mathring{X}(l_i) = X_i - m_x$,

N — number of sampled data,

λ — displacement between two ordinates $X(l)$,

m_x — mean.

The power spectrum $S_x(\omega)$ is obtained from the Fourier transform of the auto-correlation function

$$S_x(\omega) = \frac{2}{\pi} \int_0^\infty R_{xx}(\lambda) \cos \omega\lambda \, d\lambda, \tag{1.12}$$

where

$\omega = 2\pi f$ is the angle frequency (c/cm),

f – frequency (c/cm).

The auto-correlation function and power spectrum allow the separation of the periodic and random content in a profile. Determination of these values which statistically characterise a given profile completely involve certain difficulties and their use is therefore restricted to special cases.

A simple representation of the profile. In some instances a simple mathematical representation of the profile is desirable. For this purpose various representations have been suggested. These include the asperities being considered spherical and conical. Moore suggests a simple representation of the two main values when two surfaces come into contact; the number of contacts and the area of contact [125] are given by

$$N_i = C_0 \delta^n, \tag{1.13}$$

and

$$A_i = C_1 + C_2 \delta^n, \tag{1.14}$$

where

N_i is number of asperities in contact,

A_i – area of contact,

δ – distance of approach with respect to a reference plane parallel to the mean plane,

C_0, C_1, C_2, m, n are constants of the profile.

1.2 Surface processes

1.2.1 Surface tension, surface energy

Surface tension is a general property of the surface of a liquid or solid. Initially, this property was used to explain the contraction tendency of a liquid surface. A liquid drop assumes, in the absence of other forces, a spherical shape which ensures a minimum surface for a given volume. The surface tension is due to the particular position of the molecules situated at the interface. The surface layer is subjected to intermolecular forces only from the interior of the body. Reduction of the liquid surface occurs by mechanical work. Thus, the term surface tension, due to the analogy between a liquid surface and an elastic membrane, is not entirely justified. As distinguished from the tension in a membrane, surface tension does not depend on the deformation undergone by the liquid surface.

Instead of surface tension the term surface energy which is more correct from a physical standpoint is also used.

Expansion of a liquid surface implies not only mechanical work but also heat consumption if the process takes place isothermally. The total energy required for expanding the liquid surface by one square centimeter is termed the total surface energy ($\sum$). The relationship between the two quantities defined above, is

$$\sum = \sigma - T\frac{d\sigma}{dT}, \tag{1.15}$$

where

T is the absolute temperature,

σ — the surface tension.

Surface energy implies that any process of creation of new surfaces takes place by energy consumption.

Cohesion energy E_c is the work required to separate two parts of a column of material of 1 cm² cross-section. Hence, the relationship between cohesion energy and surface tension is

$$E_c = 2\sigma. \tag{1.16}$$

The above considerations are directly applicable to liquids and noncrystalline solids or to solids of undercooled liquid type. For polycrystalline solids, such as metals, the above quantities must be replaced by mean values which should take into account the existence of edges and corners.

1.2.2 **The solid-gas and solid-liquid interface**

In most practical applications the surfaces of liquids or solids constitute the interface between two phases. Even for a single substance the surface will separate the liquid or solid phase from its vapours. Obviously, in such a case the surface tension assumes values smaller than when the surface is placed in vacuum. The surface tension σ must be replaced in such a case by the solid-gas $\sigma_{s,g}$ or solid-liquid $\sigma_{s,l}$ interface tension.

Similarly to cohesion E_c, the adhesion energy E_a for the solid-liquid interface may be defined as

$$E_a = \sigma_s + \sigma_l - \sigma_{s,l}. \tag{1.17}$$

As in relation (1.17), σ_s and mainly $\sigma_{s,l}$ are difficult to determine experimentally, the wetting energy E_u is introduced as the difference between these two values

$$E_u = \sigma_s - \sigma_{s,l}. \tag{1.18}$$

Using this value, which can be determined from measurements of the liquid-solid contact angle, relation (1.17) becomes

$$E_a = E_u + \sigma_l. \tag{1.19}$$

Between the wetting energy and the total specific energy $\sum_{s,l}$ a relationship similar to (1.15) exists

$$\sum_{s,l} = E_u - T\frac{dE_u}{dT}, \tag{1.20}$$

where $\sum_{s,l}$ is the total energy resulting from the wetting of 1 cm² of solid surface, without increasing the liquid free surface. This value can be determined by measuring the heat released through the wetting of a known area of a surface. As the heat of wetting is low, of the order of 10^{-5} cal/cm², sufficiently large surface areas and accurate methods of measurement must be used. Large surfaces can be obtained only with materials in powder form. As the heat released is proportional to the area of the surface wetted the values obtained can be referred to 1 g of substance. This type of determination, although apparently simple, may lead to erroneous results. The solid material used must be thoroughly degassed. If degassing is effected by heating, the fact that some substances at a certain temperature undergo a polymorphic transformation must be considered. The purity of the liquid used must be also carefully verified. Traces of water, fatty acids or alcohols may reduce the heat of wetting of nonpolar or weakly polar liquids by up to half of the real value.

1.2.3 **Wettability of the surface**

When a solid (or liquid) surface is in contact with a liquid surface, three distinct situations are possible, depending on the contact angle. The contact angle formed between the tangents to the two surfaces at an intersection point (Figure 1.9) may by expressed as

$$\cos\alpha = \frac{\sigma_s - \sigma_{s,l}}{\sigma_l}. \tag{1.21}$$

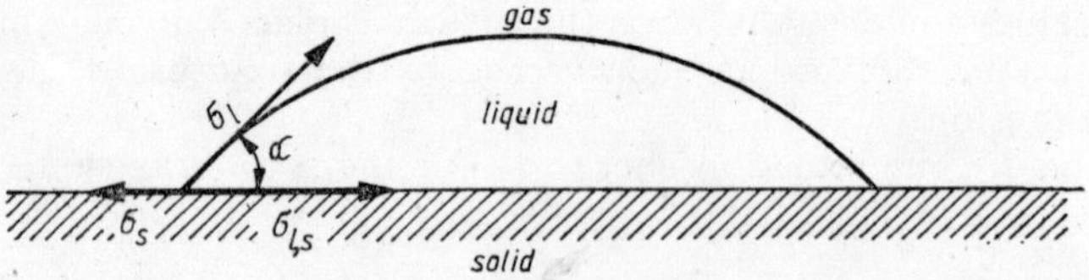

Figure 1.9 Forces acting on the surface of a liquid drop on a surface.

This relation is applicable to the case $-1 < \cos\alpha < 1$.

If $\alpha > 90°$ it may be considered that the liquid does not wet the surface; if $\alpha = 0$ the liquid spreads over the surface. Thus spreading occurs when the adhesion energy is greater than the fluid cohesion.

In the case of metals and solid alloys with a heterogeneous structure the contact angle also depends on the surface state i.e., the surface roughness, the manner of preparation, and the degree of oxidation. By using the contact angle to characterise metal surfaces it was possible to emphasize the influence of surface preparation on friction properties [94].

The measurement of the contact angle has also been applied to the examination of monomolecular layers [182]. The film state can be identified by its effect on the contact angle of a liquid drop of high surface tension. Methylene iodine is such a liquid ($\sigma = 50.8$ dyn/cm² at 20°C) of high boiling temperature and low viscosity.

The measurement of the contact angle can be made either directly through the drop or tilted plate methods, or by immersion of a cylinder [1, 171]. For measurements on surfaces subjected to frictional contact the drop method is satisfactory as surfaces of small area only are required.

1.2.4 Adsorption

Adsorption is the process which determines a modification of composition at the interface of two phases with respect to the composition in the interior of the phases. Between the excess of solute at the interface, its concentration in the solution and the surface tension, the relationship given by Gibbs is

$$\Gamma = -\frac{C}{RT}\frac{d\sigma}{dC}, \tag{1.22}$$

where

Γ is the surface excess,
C — concentration of solute,
σ — surface tension,
R — perfect gas constant,
T — absolute temperature.

If in relationship (1.22), valid for ideal solutions, C is replaced by p, that is the adsorbate partial pressure in the gaseous phase, this relationship can be also used for gas-liquid or gas-solid adsorption.

If adsorption takes place on solid surfaces, relationship (1.22), although valid in principle, cannot be used due to the difficulties experienced in the measurement of the surface tension of solids. In powder form, solid bodies allow large adsorption surfaces for a small total volume and the adsorption process can be examined by determining directly the variation of the concentration of the adsorbate or of the gaseous phase. Instead of relationship (1.22) the adsorption isotherms may be used

$$n_s = f(n_i, T), \tag{1.23}$$

for the adsorption of liquids on solids, and

$$n_s = f(p_i, T), \tag{1.24}$$

for the adsorption of gases on solids, where

n_s is adsorbate concentration on the solid surface,
n_i — adsorbate concentration in the solution,
p_i — adsorbate partial pressure.

A sensitive method for determining the adsorption isotherms uses radioactive tracers and is described in Chapter 6.

The state relations (1.23) and (1.24) can also be written in the form

$$I = f(n_s, T), \tag{1.25}$$

where I is the system enthalpy. In this way a relationship is obtained between adsorption heat, which may be determined calorimetrically, the adsorbed quantity n_s and the absolute temperature T.

Monomolecular layers, polymolecular layers. The above considerations concern phenomenological or thermodynamic aspects of the adsorption process examined by directly measured means of global quantities, such as surface tension, adsorbed quantity, or adsorption heat. Such quantities, however, do not furnish the elements required to elucidate the structure of adsorbed layers which is particularly important to explain behaviour during frictional contact. Without discussing the structure of adsorbed layers and their examination in detail, two types of layers exist:

(a) a monomolecular layer, where adsorption takes place in a single layer, the adjacent layers having the same concentration as the bulk of the gaseous or liquid phase, and

(b) a polymolecular layer which forms as a consequence of the gradual decrease of adsorption potential with distance from the surface.

1.2.5 **Chemisorption**

Adsorption occurs only when two phases are in contact at an interface. If the solid surface (solid-gas or solid-liquid interfaces) binds the adsorbed molecules only by bonds of van der Waals type, the adsorption is termed physical. With physical adsorption no exchange of electrons takes place between the molecules of the adsorbate and those of the adsorbent.

Some surfaces, such as metal, oxide, or carbon surfaces have a high degree of nonsaturation. The bonds between these surfaces and the adsorbate are chemical in nature with electron exchange. This type of adsorption is named chemisorption. A significant characteristic of chemisorption is the specific nature of the adsorbate-adsorbent bond.

A series of qualitative criteria are available for establishing the adsorption type. Nevertheless, none of these criteria is sufficient to make in all cases a distinction between physical adsorption and chemisorption. Sometimes the adsorption type remains undefined.

A first criterion to distinguish between the two adsorption types is the value of the heat of adsorption. As chemical bonds are stronger than physical bonds, the heat of chemisorption will be greater than the heat of adsorption. Typical physical adsorption values range from 1000 to 2000 cal/mol but chemisorption values range from 10000 to 100000 cal/mol [69]. Some cases of chemisorption with a thermal effect of only 3000 cal/mol have been reported [172].

Another criterion for differentiating between the two types of adsorption is the temperature range in which the process may take place. As distinguished from physical adsorption, chemisorption can also take place at temperatures much higher than the boiling point of the adsorbate. If adsorption takes place at a certain temperature and pressure (p) at which the pressure of the saturated vapours is p_0, then physical adsorption generally does not take place until the ratio p/p_0 reaches the value 0.01. This criterion cannot be considered as absolute as for some active adsorbents, particularly those with a fine porous structure, gases and vapours can be adsorbed even at values of $p/p_0 = 10^{-8}$.

Another criterion used for distinguishing chemisorption from physical adsorption is the activation energy. For a high rate of chemisorption, a certain activation energy is necessary. This may be due to the existence of a temperature threshold below which chemisorption does not take place. As physical adsorption needs no activation energy, it will take place at a certain rate at any temperature, namely at the rate at which the adsorbate reaches the solid surface. Likewise, chemisorption as distinguished from physical adsorption depends on the purity of the adsorbent surface. On the contrary, physical adsorption takes place on all surfaces.

Another difference between the two types of adsorption is the thickness of the adsorbed layer. Whilst the chemisorbed layer is always monomolecular, physically adsorbed layers may be either monomolecular or polymolecular.

The importance of chemisorbed layers in the frictional process is determined by their increased adherence to the metal surface. Also the formation of reaction layers is always preceded by chemisorption.

1.3 **Hardness**

In the contact between solid materials the hardness of the material plays an important part. Generally, hardness is defined as the resistance to penetration of a hard body, called an indenter, into the surface. Hardness is a complex quantity that cannot be deduced from the characteristics of the material obtained by tests under monoaxial stresses, although some relationships between these quantities have been established. It has been found that the hardness of steel is approximately three times the value of the yield point. Why hardness is a specific quantity irreducible to other mechanical quantities, is due to the three-dimensional loading and the effect of work-hardening during the hardness test. Thus, hardness is a quantity that characterises the behaviour of a material, or more precisely of its surface, to penetration.

Depending upon the rate at which the load is applied, the methods for hardness determination can be static or dynamic. The most popular methods are static as these provide more accurate values.

Static methods may be classified according to indenter shape and to the parameter to be measured. The Brinell method uses a spherical indenter and the indentation is measured. The Vickers method uses a square base pyramid indenter and the indentation is also measured. The Rockwell method uses a conical or spherical indenter and the depth of penetration is measured. In the first two cases hardness is expressed as the load to indentation ratio whilst the Rockwell method expresses hardness as a function of the penetration depth (Figure 1.10).

To compare the hardness measured under various loads or with different indenter sizes, the indentations should be geometrically uniform; loading must be also uniform.

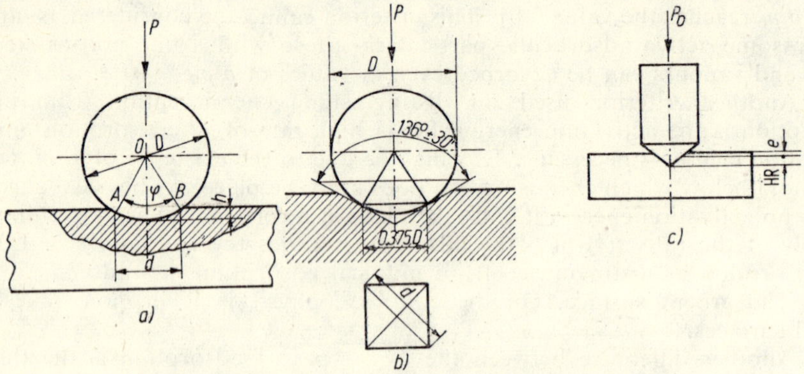

Figure 1.10 Shape of penetrators and measured quantity in: *a*, Brinell technique; *b*, Vickers technique; *c*, Rockwell technique.

With the Brinell method, the degree of loading is given by

$$k = \frac{P}{D^2}, \tag{1.26}$$

where P is the load and D the ball diameter.

The influence of ball sphericity can be minimized if the diameter d lies within the limits $0.2\,D < d < 0.5\,D$. To fulfil this condition, values of k are recommended, which give the loads to be used as a function of the indenter-diameter. In general a 15 s loading time is recommended for testing the hardness of steel and a longer time for materials with slower flow properties.

With the Vickers method the indenter is a square base pyramid with the angle between opposite faces of 136° which represents the vertex angle of a cone tangent to a sphere of diameter D at a circle of diameter $d = 0.375\,D$. In this manner the optimum condition specified above is fulfilled.

The Vickers hardness is given by the ratio of the load to the projected area of indentation. Expressed as a function of the indentation diagonal d(mm) and load P (daN), the Vickers hardness is given by

$$\mathrm{HV} = \frac{1.854}{d^2} P. \tag{1.27}$$

Using a standard indenter, the Vickers method can be used over a wide range of loads and hardness. For uniform test results, however, certain fixed loads and associated test times are chosen.

Measurement of hardness under light loads. To determine the hardness of very small surfaces, for instance the hardness of the constituents of alloys, the Vickers method uses light loads of 1 up to 500 gf.

For comparative purposes it has been agreed that the hardness determined under such loads (the microhardness) is a function of the load required to produce an indentation the diagonal of which is 5, 10 or 20 μm.

As the hardness of the constituent alloys is unknown, the indentations obtained under several loads are first measured and then the load corresponding to one value of the diagonal lengths given above is obtained by graphical interpolation.

When choosing the load account should be taken of the constituent size and hardness so that the indentation does not exceed the grain size.

Table 1.1 lists the maximum loads which may be used as a function of grain size and hardness of the metallographic constituent as well as the constituent minimum area as a function of load and hardness. The values given in Table 1.1

Table 1.1

Maximum loads and minimum dimensions inhardness determinations with microloads]*128*[

Indentation area, μm^2	Load, gf			Load gf	Minimum indentation area		
	hardness 100	hardness 400	hardness 900		hardness 100	hardness 400	hardness 900
10	0.17	0.69	1.5	1	58	14	6
50	0.85	3.4	7.7	10	582	145	64
100	1.7	6.9	15	20	1164	291	129
300	5.1	20	46	50	2910	727	323
500	8.5	34	77	100	5821	1544	646
1 000	17	68	154	Estimated hardness of			
2 000	34	137	309	some constituents of steel:			
4 000	68	274	618	ferrite			100
10 000	171	688	1 546	troostite			400
20 000	343	1 374	3 090	martensite			900

and the approximate values of the hardness of various grains minimize the number of tests.

Microhardness measurement under light loads is an important method for the examination of surface layers and the transformations which occur in the subsurface layer by frictional contact.

Rockwell hardness is determined by the depth of indentation. To eliminate zero errors the indentation depth is read under a residual load equal to that of preloading. Measured as a length, Rockwell hardness is expressed in conventional units. Correlation with Brinell and Vickers units has been established.

1.4 Relationship between surface characteristics

The solid surface, or more exactly the solid-gas or solid-liquid interface, has a complex structure and complex properties dependent upon the nature of solids, the method of surface preparation and the interaction between the surface and the environment. The methods used for the study of surfaces deal only with particular aspects under simplified conditions. The overall surface properties are obtained by the summation of specific properties although a relationship cannot always be established. The role of solid surfaces in the frictional process is determined, however, by their overall properties. Various properties of solid surfaces are discussed with emphasis on their complexity.

Topography of surfaces. A well defined geometry of the surfaces of a body is an ideal concept. Due to surface finishing processes, the surface topography deviates from the ideal form. The surface contains irregularities of various orders ranging from shape deviations to irregularities of the order of interatomic distances. No machining method, however precise, can produce molecularly flat surfaces on conventional materials. Such surfaces could not be preserved as any mechanical action would result in deterioration.

The sensitivity of surface roughness measurement methods is of the order of 0.01 μm but is reduced in stylus instruments owing to the much greater size of the tip radius. Currently, stylus instruments provide surface profiles compressed in the horizontal direction, leading to an apparent increase of asperity slopes. The real surface topography can be obtained by profilometry of surfaces along a series of sufficiently close parallel lines. In addition the ordinates of various points must be correlated so that the level lines are joined. This method, termed location profilometry, has been used to obtain maps of surfaces but owing to its complexity it has been restricted to the examination of some standard surfaces. The topography of a ground aluminium surface is shown in Figure 1.11.

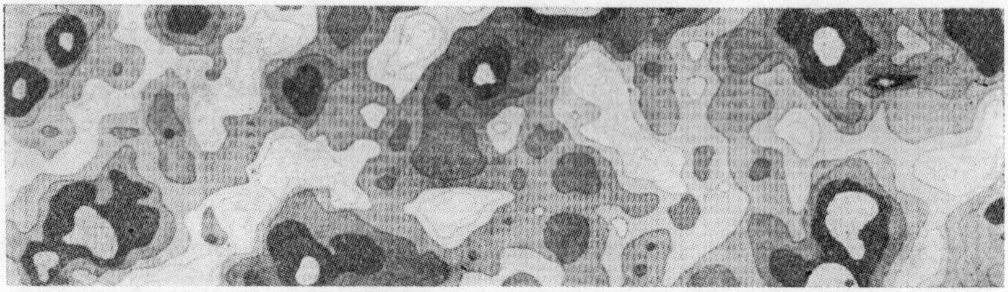

Figure 1.11 Topography of a blasted aluminium surface [180].

Structure of the surface layer of metals. Owing to the relative immobility of atoms, the properties of a solid surface depend to a great extent on the method of preparation. For example, the clean surface of a crystal obtained by cleavage will probably have a lower energy than a surface obtained by polishing. Although higher energy states, corresponding to an increased degree of lattice disorder are unstable, conversion to a more stable state is slow due to the immobility of surface atoms.

Deviations from the state of minimum energy occur within the crystalline lattice. These defects or dislocations which develop in the crystallization process play an important role in the plastic properties of the crystal. Dislocations propagate to the crystal surface, where they can be observed directly.

All metallic engineering materials are polycrystalline and in most cases include several constituents. Thus such materials may be considered isotropic only in bulk form. In polycrystalline materials, besides lattice defects, intercrystalline defects also occur. Such defects can be structural defects or inclusions. The methods of surface formation determine important changes in the distribution and density of various categories of defects in the surface layer. Also the temperature rise which unavoidably accompanies surface machining leads to phase modifications which accelerate the oxidation process.

If machining is too rough, the structural modifications may extend to a considerable depth in the material to the detriment of the subsequent service performance of the part. Machining must be controlled so that surface integrity which includes both the surface topography and the properties of the surface layer are not adversely affected.

Various forms of machining result in a series of structural modifications, which may be grouped as follows:

(a) Plastic deformation of the surface layer as a consequence of machining. This can often be recognized by the elongated form of the grains (Figure 1.12). Surface properties may be influenced by recrystallization resulting in large grains due critical cold working.

(b) Temperature rise in the surface layer during grinding may cause some phase transformations, such as the transformation of ferrite and pearlite to austenite. By quenching, austenite transforms to martensite, bainite and other products, depending on the cooling rate. Martensite is harder and more friable than the original bulk material and micro- or macrocracks due to thermal stresses may initiate in it.

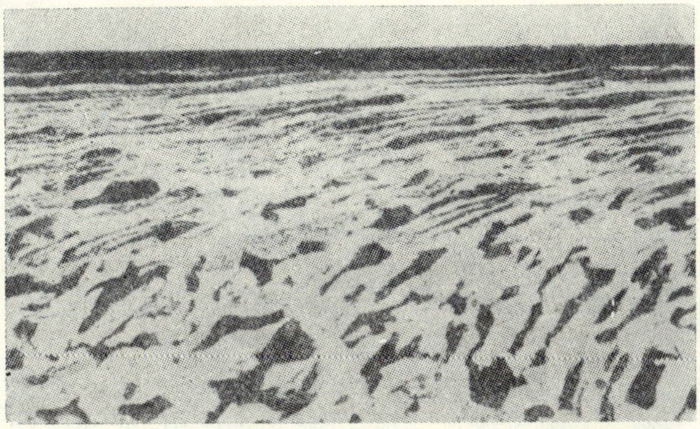

Figure 1.12 Deformation of superficial layer during surface machining [101].

(c) By plastic deformation of the surface with or without a temperature gradient, residual stresses may be released of sufficient magnitude to affect dimensional stability.

Polishing is a processing method of specific interest. Although of low intensity, affecting only a small volume of material, polishing results in the formation of an amorphous layer on the surface (Beilby layer). The amorphous character or more exactly the microcrystalline character of the layer formed by polishing was established by electron diffraction. In the case of a gold surface, manually polished, the diffraction pattern of gold appears after removing a layer of 20 μm, although traces of the process still persists down to 500 μm. The quasi-crystalline character of the layer is unstable; after some time recrystallization takes place and the diffraction pattern characteristic of the base metal reappears.

Chemical interaction with the ambient medium. With the exception of some noble metals, all metals and alloys react with oxygen and form oxides. The reaction takes place at any temperature. In most cases the metal surfaces are covered by an oxide layer the thickness of which depends on the reaction temperature and the time of reaction. The heat released by almost all processing methods increases the rate of oxidation and leads to several types of oxides. For instance, with a steel surface the temperature reached during frictional contact influences the composition and structure of the oxide layer (Figure 5.1).

The less intense the method of surface processing, the closer are the properties of the surface layer to the properties of the bulk of the material, and the thinner is the oxide layer.

A method such as wet polishing produces surfaces covered by a thin Beilby type layer and initially by thin oxide layer. Such surfaces provide reproducible friction properties [94].

During the friction process, due to the rise in temperature chemical reaction with the ambient medium is accelerated. When a metal friction pair operates in air the process may take place between the oxide layers of the two surfaces (mild wear regime). The presence of a lubricant and additives causes the formation of solid reaction layers which are important in surface protection.

The relative importance of various layers is shown schematically in Figure 1.13.

During frictional contact the structure of the surface layer undergoes modifications in thickness and constitution. Generally, during running-in, the frictional properties of surfaces improve. Although the running-in of surfaces usually consists of material removal a smoothing of the surface is possible by plastic deformation.

The solid surface and its role in physical processes is a complex phenomenon which can only partially be described by quantities such as roughness, hardness and surface energy. Surface changes generally imply modification of all

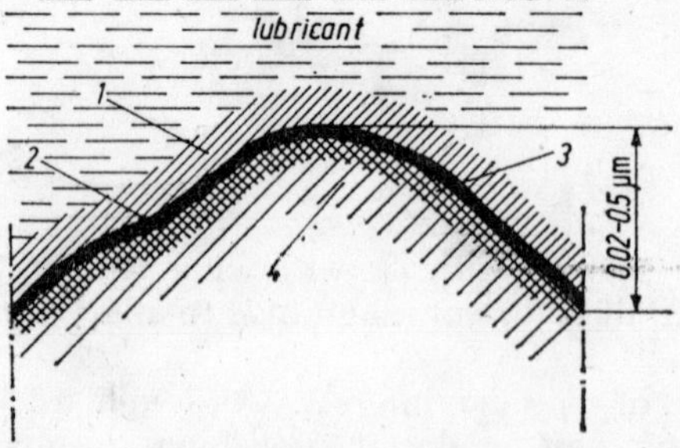

Figure 1.13 Structure of superficial layer (orientative dimensions): 1, monomolecular adsorbed layer (~30Å); 2, oxide layer (20 ÷ 100 Å); 3, microcrystalline metallic layer (50 ÷ 500 Å); 4, deformed layer (~ 1 μm).

properties. A thorough knowledge of surface properties is of particular importance in friction and wear processes, fatigue and other phenomena by which machine parts deteriorate in service.

2 Contact between solids

2.1 Real and apparent area of contact

Considering the microgeometry of solid surfaces and their properties it may be stated that the real area of contact is much smaller than the apparent area of contact and that the real area of contact increases with load.

For two conformal surfaces in contact (Figure 2.1) the following types of contact area can be defined:

(a) the apparent contact area A_a,

(b) the real contact area A; on the real area the bodies approach up to distances in which atomic or molecular forces act,

(c) the area of the contact spots A_p, with two conformal surfaces, due to second order irregularities, the contact microsurfaces are generally grouped into contact spots.

Calculation of the real contact area and its dependence on load can be made if the topography and mechanical properties of the solid surface are considered. As the friction force is determined by surface adhesion, various attempts to calculate the real contact area should have led to its linear dependence on load in confirmation of the Amonton-Coulomb law.

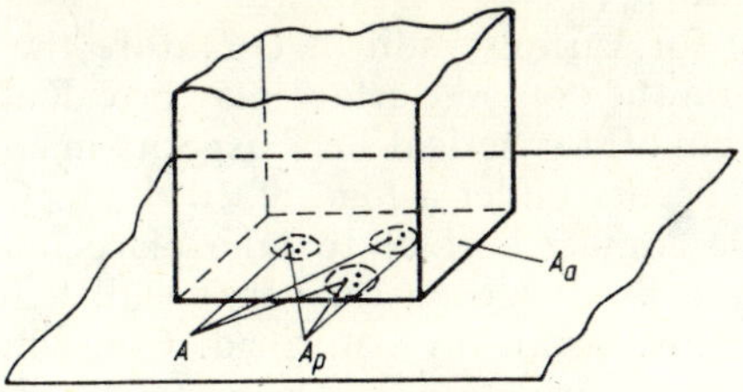

Figure 2.1 Contact between two plane surfaces.

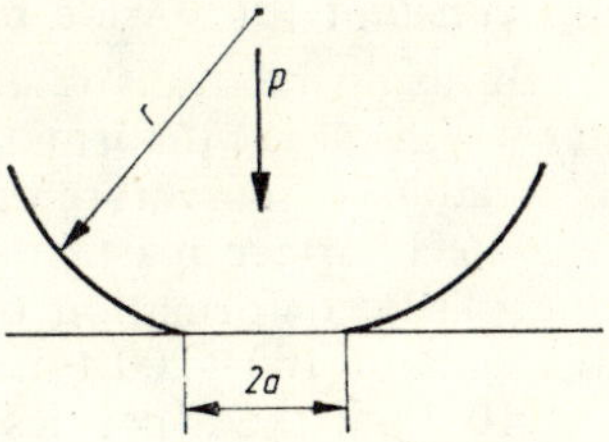

Figure 2.2 Elastic contact between a sphere and a plane.

In the case of an elastic contact between a sphere of material 1 and a perfectly flat surface of material 2, the radius of the circular contact area a is (Figure 2.2)

$$a^3 = \frac{3}{4} \cdot \frac{Pr}{E'}, \qquad (2.1)$$

where

P is the load,

r — the sphere radius and $\frac{1}{E'} = \frac{1-\nu_1}{E_1} + \frac{1-\nu_2}{E_2}$,

ν — Poisson's ratio.

The contact area A is

$$A = kP^{2/3}. \tag{2.2}$$

2.2 Plastic contact

To obtain a linear dependence of the real area of contact on load, Holm [84 and Bowden and Tabor [39] assumed that the asperities in the contact areas are plastically deformed. For the contact shown in Figure 2.2, the elastic limit is reached at a point in the softer body when the mean contact pressure p_m reaches the value

$$p_m = 1.1\,\sigma_e, \tag{2.3}$$

where σ_e is the elastic limit for the body of lower hardness, and

$$p_m = \frac{P}{\pi a^2}. \tag{2.4}$$

For plastic deformation to occur over the whole surface of contact it is necessary that the pressure $p_m \simeq 3\sigma_e$; this value is also termed yield pressure in hardness tests (σ_c'). It has been confirmed experimentally that this quantity remains constant for a wide range of loads.

By using relations (2.1) and (2.3) for various radii of curvature the load under which plastic deformation occurs on the contact surface may be calculated. For a radius of curvature of $r = 10^{-4}$ cm of a spherical steel asperity in contact with a steel surface plastic deformation occurs under a load of 10^{-3} gm. Considering plastic deformation of the whole contact surface to take place under a load of about 100—200 times greater means that even for very small loads of about 10^{-1} gm the asperities in contact are plastically deformed. Consequently Bowden and Tabor [39] assumed plastic deformation of the asperities in the contact area and suggested the following simple relation for the real area of contact

$$A = \frac{P}{\sigma_c'}. \tag{2.5}$$

The considerations used for deducing relation (2.5) assumed both the absence of any work-hardening effect and the extreme of spherical asperities harder than the material of the counterpart. It can be shown that for materials which work-harden the area of contact is a function of load

$$A = k \cdot P^n,$$

where n varies within the range 0.8—1, depending upon the material (the higher value is for materials which do not work-harden). The shape of asperities influences only the value of k but not the exponent n.

Taking account of the fact that by the very process of surface machining the asperities are considerably deformed the theory of plastic contact according to relation (2.5) is considered valid for all cases.

2.3 Effect of tangential force

Using relation (2.5) and assuming that frictional force is a result of an adhesion process in the real area of contact, the friction coefficient μ may be obtained from the simple expression

$$\mu = \frac{\tau}{\sigma_c'}, \qquad (2.6)$$

where τ is the shear strength of the softer metal.

If the metallic materials of the bodies in contact are ideally plastic then $\sigma_c' \simeq 5\tau$ and a friction coefficient of $\mu \simeq 0.2$ results, which is lower than the friction coefficient determined experimentally for dry surfaces. For clean metallic surfaces $\mu \simeq 1$. This disagreement is accounted for in the theory of plastic contact by increase of the real contact surface, due to the existence of a tangential force in the friction process.

In a friction pair, in addition to a normal force, a tangential force also arises as a result of the relative motion of the mating parts. This tangential force

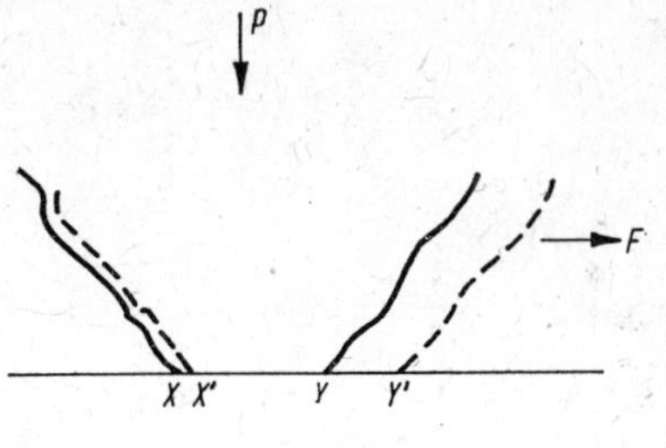

Figure 2.3 Effect of tangential force on the area of contact.

tends to increase the real area of contact contributing to surface plastic deformation.

In the two-dimensional case (Figure 2.3) for plastic contact between the initial normal stress σ_0 and the stress σ, in the presence of a tangential force the relationship is

$$\sigma^2 + 3\tau^2 = \sigma_0^2. \qquad (2.7)$$

In the three-dimensional case the same relationship is assumed valid if another coefficient is taken for the tangential stress τ

$$\sigma^2 + \alpha\tau = \sigma_0^2, \tag{2.8}$$

where α is a function of the surface material.

Equation (2.8) can be written in a form such that the area of surface contact explicitly appears before and after the application of a tangential force

$$1 + \alpha\Phi^2 = \left(\frac{A}{A_0}\right)^2, \tag{2.9}$$

$$\Phi = \frac{F}{P},$$

where A_0 is the real area of contact in the absence of the tangential force, F is the tangential force and P the load.

This relation is valid in the absence of motion as well as during displacement when $\Phi = \mu$. Increase of the real area of surface contact has been determined by measuring the coefficient Φ by electrical contact resistance as a function

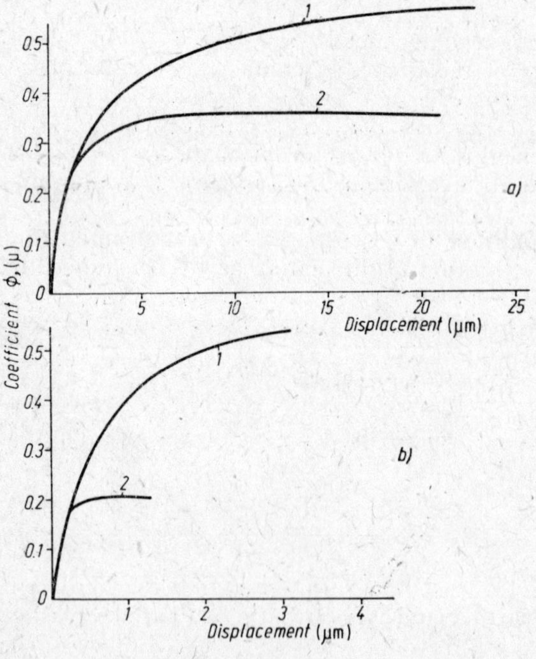

Figure 2.4 Coefficient Φ, (μ) versus displacement [50]: *a*, platinum; *b*, steel; **1**, clean surfaces; **2**, surfaces lubricated with saturated solution of lauric acid in cetane.

of displacement. Figure 2.4 which shows the value of Φ as a function of the tangential displacement, indicates how the surface area of contact increases. When Φ assumes the value of the static coefficient of friction the displacement is permanent.

2.4 Elastic contact

Although the assumption of plastic contact leads to a satisfactory interpretation of the Amonton-Coulomb law, it contradicts other experimental facts. Archard 10] takes into account the mechanism of dry friction and wear and concludes

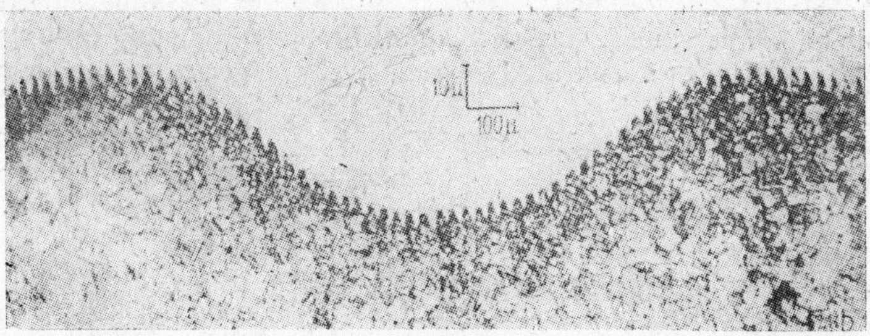

Figure 2.5 Plastic deformation of a rough surface [39].

that repeated plastic deformation of asperities is not possible. Plastic deformation may occur during the first passage of the moving part; subsequently, however, deformation of the asperities in contact should be elastic as for the separation of a wear particle a considerable number of passages is necessary.

The resistance of asperities to plastic deformation can be observed by microscopical examination (Figure 2.5). Even when the yield point of the bulk material is exceeded, the asperities still retain their initial form. Clearly, by repeatedly subjecting the surface shown in Figure 2.5 to loads lighter than the initial load, the asperities will undergo only elastic deformation.

Some authors [10, 99] reported an almost linear dependence of the real area of contact on load can be obtained if the assumption of the multiple elastic contact is used.

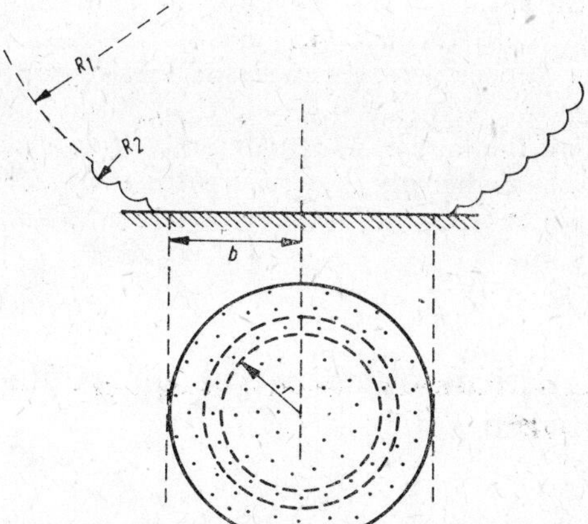

Figure 2.6 Multiple contact model (Archard).

Considering the contact between a spherical surface of radius R_1, with uniformly distributed spherical asperities of radius R_2, and a flat surface, a load exponent close to unity can be obtained (Figure 2.6).

Assume that the sphere of radius R_1 is smooth, the load carried by a ring of radius r and width dr, for elastic deformation, is

$$\mathrm{d}P = \frac{2}{3}\frac{P^{1/3}}{k_1}\left(1 - \frac{r^2}{b^2}\right)^{1/2} 2\pi\, r\, \mathrm{d}r, \tag{2.10}$$

where P is the load and b is the radius of the contact spot.

If $R_2 << R_1$, the presence of asperities of radius R_2, uniformly distributed with density m on the sphere surface do not influence the state of stresses in the sphere of radius R_1. Then, the load carried by an asperity in the ring of radius r is

$$p = \frac{\mathrm{d}P}{q} = \frac{2P^{1/3}}{3mk_1}\left(1 - \frac{r^2}{b^2}\right)^{1/2}, \tag{2.11}$$

where $q = 2m\pi r dr$ is the number of asperities in the ring of radius r.

Since the deformation is elastic the contact area of an asperity is

$$a = kp^{2/3} = k\left(\frac{2}{3m\, k_1}\right)^{2/3} P^{2/9}\left(1 - \frac{r^2}{b^2}\right)^{1/3} \tag{2.12}$$

and the contact area on the ring of radius r will be qa.

The total contact area is

$$A = \int_{r=0}^{r=b} qa. \tag{2.13}$$

Integrating equation (2.13) gives

$$A = k_2 P^{8/9}, \tag{2.14}$$

that is an almost linear dependence of the area of contact on load is obtained although the deformation is elastic.

By further dividing the asperities of radius R_2 into asperities of radius R_3, where $R_3 << R_2$, that is an increase of the number of contacts, leads to a further increase of the load exponent without it exceeding unity. This behaviour is due to the fact that when the load increases the total area of contact increases, mainly due to increase in the number of contacts and not by increase of each asperity area of contact.

2.5 Statistical treatment of the contact between solids

Although the models of elastic and plastic contact provide some agreement with experimental results, both use assumptions which partially depart from actuality. An attempt to take account of the statistical surface topography has been made by Greenwood and Williamson [79, 180]. They used for characterising the surface topography quantities experimentally determined, namely the distribution of asperity heights Φ, the asperity mean curvature radius β and the standard deviation of the distribution of the asperity heights σ.

For a spherical asperity of radius β, for elastic contact, the radius of the contact surface r, the contact area and the pressure p are given by

$$r = \beta^{1/2}w^{1/2}; \; a = \pi\beta w; \; p = \frac{4}{3} E'\beta^{1/2}w^{3/2}, \tag{2.15}$$

where w is the approach of bodies as a result of deformation.

Figure 2.7 Contact between a plane rough surface and a smooth plane surface.

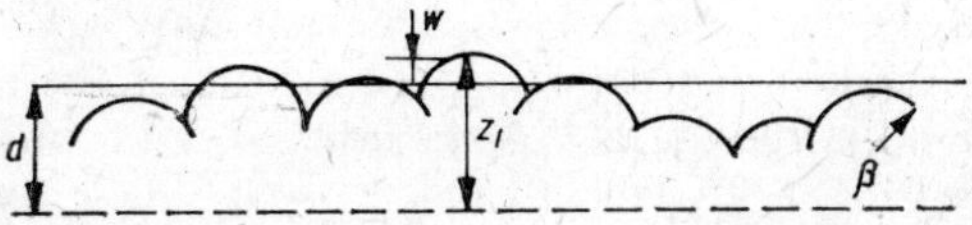

Consider the elastic contact between two flat surfaces one rough and covered with spherical asperities, of radius of curvature β and the other nominally flat. All asperities of height z greater than d will come into contact (Figure 2.7).

The probability that an asperity of height z is greater than d, that is in contact, is

$$\text{prob}\,(z > d) = \int_d^\infty \Phi(z)\,\mathrm{d}z. \tag{2.16}$$

Since $w = z - d$ and $a = \pi\beta w$, and the total number of asperities is N, then the real area of contact is given by

$$A = \pi N\beta \int_d^\infty (z - d)\Phi(z)\,\mathrm{d}z. \tag{2.17}$$

Introducing the standard deviation σ of the asperity heights, the density of asperities η and the apparent area of contact A_a gives for the real area of contact

$$A = \pi\eta A_a\beta\sigma F_1(h), \tag{2.18}$$

where

$$F_1(h) = \int_h^\infty (s - h)\,\Phi(s)\,\mathrm{d}s, \tag{2.19}$$

in which

$$h = \frac{d}{\sigma}.$$

Similarly expressions for the number of contact points n and for load P, can be obtained.

To resolve relation (2.19) a certain distribution of asperity heights must be considered. In the case of an exponential distribution of asperities $\Phi^*(s) = e^{-s}$, the number of contact points, the real area of contact and the pressing force are

$$n = \eta A_a e^{-h}, \tag{2.20}$$

$$A = \pi\eta\beta\sigma A_a e^{-h}, \tag{2.21}$$

$$P = \pi^{1/2}\eta\beta\sigma E'(\sigma/\beta)^{1/2} A_a e^{-h}. \tag{2.22}$$

From relations (2.20) — (2.22) a proportionality is obtained between the number of contacts and load, and between load and real area of contact. It may thus be observed that the mean area of contact points depends linearly on load, and the mean pressure on the contact surface is independent of it.

For a Gaussian distribution, only approximate solutions can be obtained for equation (2.19). The dependence of the real area of contact and of pressure on load is shown in Figures 2.8 and 2.9 for the values $\eta = 300/\text{mm}^2$, $\beta\sigma = 10^{-4}\text{mm}^2$, $E'(\sigma/\beta)^{1/2} = 25$ daN/mm².

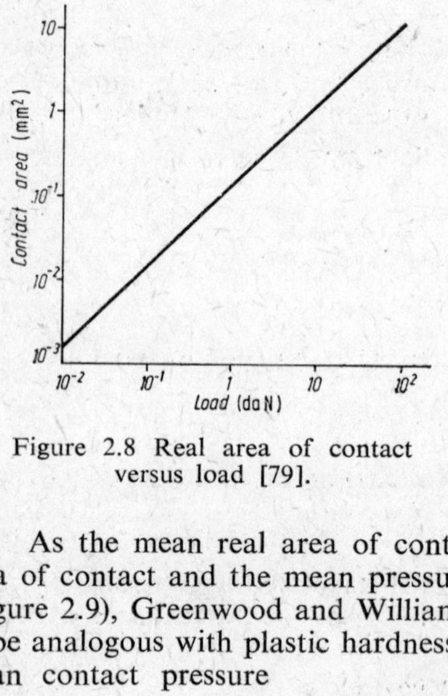

Figure 2.8 Real area of contact versus load [79].

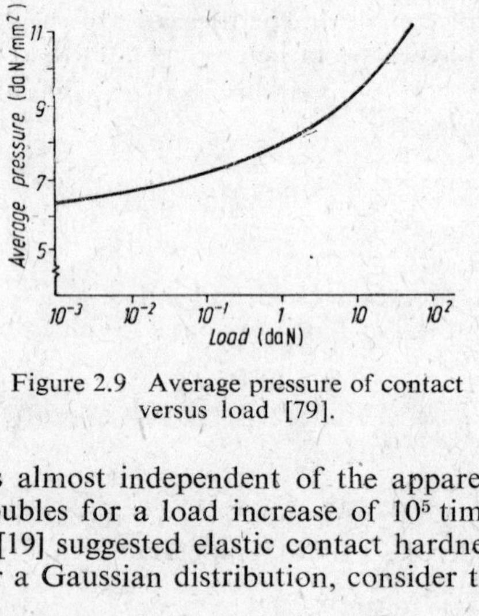

Figure 2.9 Average pressure of contact versus load [79].

As the mean real area of contact is almost independent of the apparent area of contact and the mean pressure doubles for a load increase of 10^5 times (Figure 2.9), Greenwood and Williamson [19] suggested elastic contact hardness to be analogous with plastic hardness. For a Gaussian distribution, consider the mean contact pressure

$$\frac{P}{A} = 0.25\, E'(\sigma/\beta)^{1/2}, \tag{2.23}$$

which allows the real elastic area of contact to be calculated if σ and β are known.

In order to separate the elastic from the plastic regime of deformation, the same authors introduced the quantity

$$\Psi = \frac{E'}{H}\sqrt{\frac{\sigma}{\beta}}, \tag{2.24}$$

where H is the hardness of the softer material.

The quantity ψ, termed the plasticity coefficient, involves the properties of the materials and the geometry of the surfaces in contact and permits the determination of the critical load under which an elastic deformation becomes a plastic deformation. For values $\psi < 0.6$ deformation is elastic up to very heavy loads; for $\psi > 1$ deformation is entirely plastic, even under light loads. An important factor in the value of ψ is the surface roughness, characterized by σ and β. Most technical surfaces obtained by the usual processing methods have a plasticity coefficient higher than unity (Figure 2.10), that is the deformation of the contact surface is mainly plastic.

However, surfaces exist for which the plasticity index is smaller than unity. For instance, in the case of a bearing ball with characteristics $R_a = 0.04$ µm, standard deviation $\sigma = 0.024$ µm and a mean radius of asperities $\beta = 150$ µm, the resulting plasticity coefficient is $\psi = 0.25$. This very low value is due to the great hardness of the ball material. By polishing a plasticity coefficient $\psi < 1$

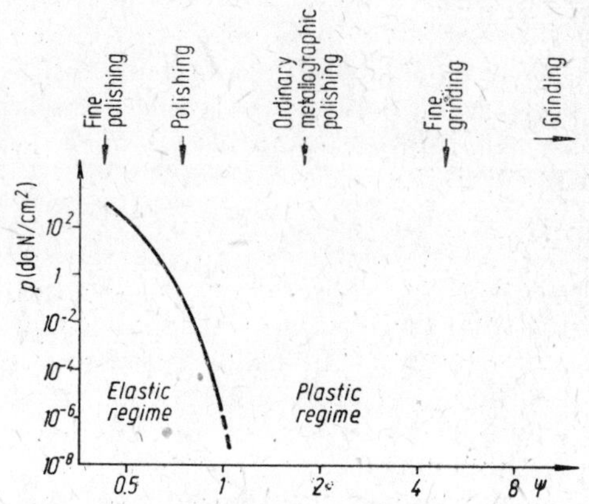

Figure 2.10 Deformation of surface versus plasticity index and apparent pressure p for the contact between two plane surfaces [166].

was obtained for less hard surfaces. For example, for a polished surface of unhardened steel of $R_a = 0.012$ µm, $\beta = 0.5$ mm and $\sigma = 100$ Å a plasticity coefficient of $\psi = 0.3$ was obtained.

By polishing or running-in, the plasticity coefficient decreases; this is in agreement with the general evolution of the rate of wear. Decrease in the wear rate corresponds to the transition from plastic to elastic contact.

2.6 Experimental determination of the real contact area

A method for determining the real area of contact between two metallic surfaces or other materials with low electrical resistivity consists of the measurement of the electrical contact resistance.

If a current flows through two massive conductors in contact, the resistance caused by the sudden decrease of the cross-section (since $A << A_a$) is termed the constriction resistance. When the whole contact surface reduces to a circle of radius a, then the constriction resistance R_c is given by

$$R_c = \frac{\rho}{2a}, \tag{2.25}$$

fwhere ρ is the conductor specific resistivity, and a the radius of the real area of contact.

If contact takes place through a layer of given constant thickness and has the resistance σ (Ω cm^2), then the contact resistance is

$$R = (\rho/2a) + \frac{\sigma}{a^2}\cdot \tag{2.26}$$

The real area of contact is distributed over the entire apparent contact surface. If each contact surface, assumed to be a circle of radii $a_1, a_2, \ldots$, is sufficiently far from the other to avoid interference in reciprocating contact, then the total constriction resistance is

$$R_c = \frac{\rho}{2\sum a_i}, \tag{2.27}$$

i.e., the total constriction resistance is equal to the local constriction resistance mounted in parallel. Consider the mean value of the individual contact radius $\bar{a}$, then the sum in relation (2.27) is equal to na. By measuring the contact resistance in vacuum in the absence of an oxide layer the quantity a or $n\bar{a}$ in relation (2.25) can be determined.

In the former case the radius of the contact surface for the point contact was obtained directly. In the latter case it is also necessary to determine the number of contacts. To determine the value of a, Holm [84] used a copper-graphite pair for which he measured the contact time on a small portion of a friction surface. From the probability of the existence of contacts on the surface measured he deduced the average number of contacts.

3 The temperature of friction surfaces

Most of the energy lost because of friction is released as heat. Friction between solids causes heat due to the elastic and plastic deformation of the surface layer of the materials. Thus a high temperature gradient occurs in the direction normal to the surface. The experimental determination of such a temperature gradient close to the surface is difficult.

The first attempts to derive a theoretical solution to the problem of the surface temperature of solids were made by Bussinesq, as far back as the past century. Further research covering the friction of solids was carried out by Blok [33], Archard [11] and Cameron [203].

3.1 Surface temperature for linear contact

The general heat conductance differential equation for the three dimensional case is

$$\nabla^2\theta = \frac{\lambda}{\rho c} \cdot \frac{\partial \theta}{\partial t} \tag{3.1}$$

where

θ is the temperature,
λ — the conductivity,
ρ — the specific weight,
c — the specific heat capacity,
t — the time.

To simplify the problem, assumptions are made in order to reduce it to the one dimensional conduction case. For this it is assumed that the displacement speed is sufficiently high to allow heat transfer especially in the normal direction to the surface. The heat source is also assumed to be one dimensional.

Under these conditions, equation (3.1) becomes

$$\frac{\partial^2 \theta}{\partial^2 y} = \frac{\lambda}{\rho c} \cdot \frac{\partial \theta}{\partial t}. \tag{3.2}$$

The solution of this equation

$$\theta = \frac{Q}{\sqrt{\pi\lambda\rho c t}}\, e^{-\frac{y^2}{\frac{4\lambda}{\rho c}t}}, \tag{3.3}$$

where

Q is the source strength,

t — the time elapsed from the heat release.

However, what is of interest in this case is the surface temperature, i.e. $y = 0$ and hence

$$\theta = \frac{Q}{\sqrt{\pi\lambda\rho c t}}\,. \tag{3.4}$$

In practice, linear sources are of the strip type, such as in toothed wheels and it may be assumed that the strip width is much smaller than the strip length.

To calculate the maximum surface temperature of a strip type source which displaces with velocity V, a $q = q(\zeta)$ distribution of the heat flow intensity is assumed (Figure 3.1).

The temperature located in the direction x due to a $q(\zeta)\,d\zeta$ source of dt duration is [169]

$$d\theta = \frac{q(\zeta)\,dt}{\sqrt{\pi\lambda\rho c t}}\,. \tag{3.5}$$

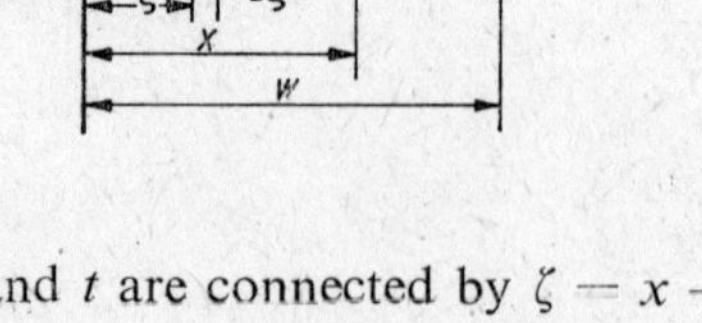

Figure 3.1 Strip type source of heat.

Since ζ and t are connected by $\zeta - x - tV$, equation (3.5) becomes

$$d\theta = -\frac{1}{\sqrt{\pi\lambda\rho c V}}\frac{q(\zeta)\,d\zeta}{\sqrt{x-\xi}}, \tag{3.6}$$

or

$$\theta = -\frac{1}{\sqrt{\pi\lambda\rho c V}}\int_0^x \frac{q(\zeta)\,d\zeta}{\sqrt{x-\zeta}}\,. \tag{3.7}$$

Equation (3.7) gives the temperature distribution on the surface of a solid, considering a given distribution of the heat flow intensity. The maximum surface temperature is of particular interest and is obtained by differentiating equation (3.7) with respect to x.

The maximum surface temperature of a strip type source θ_{max} is given with a good approximation by equation [33]

$$\theta_{max} = A \frac{q_m}{\sqrt{\lambda \rho c}} \sqrt{\frac{w}{V}}, \tag{3.8}$$

where A is a constant which depends upon the heat flow distribution type. For an ellipse shaped distribution $A = 1.13$; for a constant intensity $A = 1.11$;

q_m — medium heat flow intensity ($\text{cal.cm}^{-2}\text{s}^{-1}$),

w — strip width,

V — speed of the source.

In the case of friction between solids, the two bodies are generating heat and frictional heat is distributed in both solids. To calculate the distribution ratio one postulates equal temperature for the two contacting surfaces. In order to make use of equation (3.8) for the second body, the latter should be in motion with regard to the heat source, i.e. with the conjunction area. Both V_1 and V_2 should be high enough to fulfil the conditions given by (3.2).

Because of the equality maximum of temperatures of the contact surfaces (the distribution of the heat flow intensity being the same for both solids) $(\theta_{max})_1 = (\theta_{max})_2$ and

$$\frac{q_{m1}}{\sqrt{\lambda_1 \rho_1 c_1}} \sqrt{\frac{w}{V_1}} = \frac{q_{m2}}{\sqrt{\lambda_2 \rho_2 c_2}} \sqrt{\frac{w}{V_2}}. \tag{3.9}$$

Taking into account that

$$q_{m1} + q_{m2} = q_m, \tag{3.10}$$

both equations (3.9) and (3.10) give

$$\theta_{max} = A \frac{q_m}{\sqrt{\lambda_1 \rho_1 c_1} \sqrt{V_1} + \sqrt{\lambda_2 \rho_2 c_2} \sqrt{V_2}} \sqrt{w}, \tag{3.11}$$

where indices 1 and 2 refer to both solids.

3.2 Surface temperature for point contact

A simple method for calculating the surface temperature for the point contact (i.e. circular contact with small radius) has been suggested by Archard [11].

Consider the contact surface between two solids B and C of circular form of radius a. Heat develops as solid B travels with a speed V (Figure 3.2). Solid B receives heat from a motionless source while solid C received heat from a moving source.

The heat developed is distributed between both surfaces so that the medium temperature θ_m at the interface should be the same for both solids. The θ_m temperature is considered in comparison with an internal point of the solid well removed from the surface; the temperature of the solid is considered to be zero.

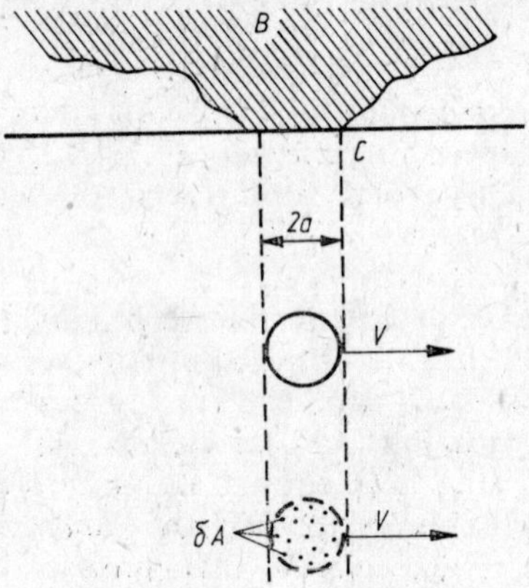

Figure 3.2 Pointlike heat source.

If it is assumed that the heat distributed in the two solids is known, the temperature of solid B from a motionless source may be written by analogy with the constriction resistance [84]

$$\theta_m = \frac{Q_B}{4a\lambda_B}, \tag{3.12}$$

where

Q_B is power of the heat source for solid B,
λ_B — thermal conductivity of solid B,
a — contact radius.

For speeds of V sufficiently low, when it may be considered that in any position there is sufficient time for the stationary regime to be achieved, the temperature of solid C may be calculated using the same equation

$$\theta_m = \frac{Q_C}{4a\lambda_C}. \tag{3.13}$$

To mark the quantitative speed limit for which equation (3.12) and (3.13) are no longer valid, the value

$$L = \frac{Va}{2\chi}, \tag{3.14}$$

is introduced, where $\chi = \lambda/\rho c$ is the thermal diffusivity.

The value L represents the ratio between the time required to allow the temperature, in the case of an instantaneous source, to reach the maximum value to a depth equal to a and the time during which the point of the corresponding surface is subject to the heat source.

When $L < 0.1$ equation (3.13) may be used but when $L > 5$, equation

$$\theta_m = \frac{0.31\, Q_C}{\lambda_C a} \left(\frac{\chi_c}{Va} \right)^{1/2}, \tag{3.15}$$

is valid. For the domain $0.1 < L < 5$ equation

$$\theta_m = \alpha \frac{Q_C}{4a\lambda_C}, \tag{3.16}$$

is valid, where $\alpha = 0.85 \div 0.35$ when $L = 0.1 \div 5$.

Q_C and Q_B, i.e. the quantities of heat passing into solids C and B are unknown, so to determine the average surface temperature, which is the same for both solids, Archard [11] suggests calculation of the temperature of the motionless solid surface Q_C by means of equation (3.12) assuming that the total heat passes into solid B; then calculation of the surface temperature of solid C by means of equation (3.15) or (3.16) according to the value of the parameter L, assuming that the total heat passes into solid C. The real temperature of the contact surface is given by the equation

$$\frac{1}{\theta_m} = \frac{1}{\theta_B} + \frac{1}{\theta_C}. \tag{3.17}$$

To calculate the surface temperature by relations (3.12) $\div$ (3.16), it is necessary to know the real area of surface contact between the solids. Lack of precision in defining the real area of contact also affects the value of the surface temperature. The lack of precision increases with dry friction between metals, because an oxide layer is present.

Using equations (2.1) and (2.5) to determine the area of contact between a sphere and a plane of similar material, it is possible to deduce a series of equations for very low speeds when $Q_B = Q_C = 0.5\, Q$ or for very high speeds ($L > 100$) when $Q_C = Q$.

For instance, with a plastic contact between a sphere and a plane there are the equations:
when $L < 0.1$

$$\theta_m = \frac{\mu g}{J} \frac{(\pi \sigma_c')^{1/2}}{8\lambda} P^{1/2} V, \tag{3.18}$$

or when $L > 100$

$$\theta_m = \frac{\mu g}{3.25\, J} \frac{(\pi \sigma_c')^{3/4}}{(\lambda \rho c)^{1/2}} P^{1/4} V^{1/2}, \tag{3.19}$$

where

σ_c' is yield pressure in hardness tests,
P — load,
J — conversion constant,
g — acceleration due to gravity,
μ — coefficient of friction.

For most applications, the maximum temperature affects the processes developing on the friction surface. To determine the maximum temperature it is necessary to utilize the minimum contact areas when dealing with equations (3.12) and (3.13).

Relations (3.18) and (3.19) assume that the real area of contact is equal to the apparent area of contact. Maintaining the analogy with the electrical contact resistance, it is possible to show that the resulting equations are still reasonably valid also when the real points of contact are distributed within the apparent contact surface (Figure 3.2).

The conditions represented by diagram (3.2) are only for solid B moving and solid C motionless. When both solids are moving, as in the case of gears, equation (3.15) or (3.16) should be used to calculate both θ_B and θ_C, taking into account the sliding speed in order to determine the heat developed.

3.3 Experimental determination of surface temperature

Experimental determination of the surface temperature achieved by friction is difficult owing to the high value of the temperature gradient to the surface. For instance, equation (3.3) shows that when t = constant, the temperature normal to the surface has an exponential variation, the distance from the surface being a second power index.

Considering the temperature distribution in the near surface zone, it is obvious that the usual temperature transducers, thermocouples and resistance thermometer, even if miniaturized are still too large to allow accurate measurement of the surface temperature.

To measure the temperature of the actual friction surface, use is made of the thermoelectric effect in metals and the variation of electroconductivity with temperature. To achieve this the active volume of the transducer must contact the friction surface. Methods of doing this introduce several factors deleterious to the precision of the measurements.

Dynamic thermocouple method. This technique used a natural thermocouple [39, 61], the two elements of the couple being the moving units, with the junction the real contact surface. The voltage developed by the thermocouple produces a characteristic image on a cathode ray oscillograph. Disregarding the absolute temperature which depends on speed and load, the existence of temperature peaks, superposed upon a predominant mean temperature may be observed. The duration of the temperature peaks, which is of some tens of milliseconds, as well as their amplitude indicate that the contact surface temperature under dry friction at contact fluctuates. The maximum temperature develops when the area of surface contact is a minimum. Use of this technique has indicated maximum temperatures up to 900 — 1000°C for steel surfaces [39].

Using a friction couple of a ball and a rotating cylinder, Furey [61] compared the values of the mean temperatures of the surface calculated according to equations (3.12) — (3.16) with the measured mean value (Table 3.1). Significant differences may be noticed due to either the use of too small an area of surface contact or to intrinsic errors in the dynamic thermocouple method.

If it is considered that by frictional force the real area of surface contact increases, an improved correlation is obtained between the calculated and the measured values. To calculate the contact area during motion, Dayson [53] considered the equation which provides an increased area of surface contact due to the tangential force (2.9). Replacing Φ by the friction coefficient gives

$$1 + \alpha\mu^2 = \left(\frac{A}{A_0}\right)^2. \tag{3.20}$$

Measuring μ and assuming that the contact surface remains circular after deformation, it is possible to determine a corrected a value which enters into equations (3.12) — (3.16). The α coefficient is known only for a few pure metals; however, it has been ascertained that over a certain value, α has no significant

Table 3.1

Theoretical and experimental values of surface temperature increase C°[53]

Load (gf)	Speed rot/min	Theoretical value: plastic contact	Theoretical value: elastic contact	Experimental value (Furey)[61]	Corrected value
30	60	18.4	16,1	5.6	5.2
	240	65	57.8	16.1	16.2
	480	251	226	27	39
	960	446	390	63	59
60	60	45	45	13.1	9.1
	240	109	109	22.8	22
	480	335	335	52	49
	960	577	577	73	73
250	60	25	33	14.4	9.9
	240	99	133	48	28
	480	195	264	71	54
	960	387	537	90	86

influence upon the results. Calculation of the average surface temperature by taking into consideration the increase in the area of surface contact as shown in the last column of Table 3.1 gave results much closer to the experimental results.

Welding onto the friction surface (Figure 3.3) is another method of making close contact with the thermocouple. In this method only one of the friction couple components is used for the thermocouple together with a thin insulated wire which is welded to the surface, by for instance, discharge of a condenser. Although part of the junction is inevitably located below the friction surface, the technique has the advantage that it can be used with couples of the same material. The values measured by this method are mean values. The precision is good because of the welded junction.

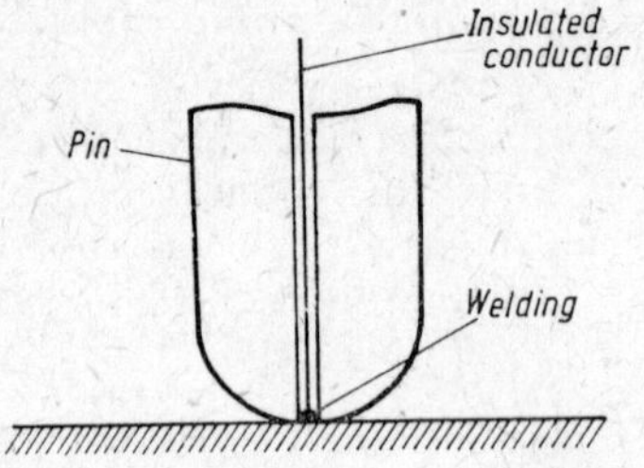

Figure 3.3 Function of thermocouple on the friction surface.

Resistance thermometer. Measurement of surface temperature by a resistance thermometer is possible in certain cases, by use of the thin film resistance method. By a suitable technique a few microns thick metal film is applied to a small isolated area of the contacting surface of one member of the friction couple, so that the terminals can be taken outside the zone under stress. The resistance of the film which depends on the temperature may be measured by an electronic bridge. This type of thermometer is subject to deterioration by wear of the member of the friction couple.

Trailing thermocouple. Another useful technique uses a thermocouple in contact with the surface of one member of the friction couple. This technique, the principle of which is shown in Figure 3.4, does not necessitate modification of the friction couple. The comparison between the real surface temperature, measured by a thermocouple the active part of which should be located approximately 0.01 mm below the surface and that measured by a contact thermocouple under various compressive forces without lubrication is shown in Figure 3.5. In the absence of a lubricant the temperature measured depends on the compressive force, although under heavier loads the influence on precision is of little significance.

When a lubricant is present the temperature measured by the contact thermocouple is very near the real temperature for large scale compressive forces and sliding speeds (Figure 3.6). A disadvantage of the contact thermocouple method is that the thermocouple gives a mean value between the real surface temperature and the temperature of the lubricant boundary layer. Under steady state conditions the two temperatures are similar. When the lubricant temperature alters suddenly, a difference between the two temperatures is found. This aspect is counterbalanced by the advantage of the thermocouple not being adversely affected by wear of the friction couple.

Infrared radiation measurement. With friction couples where one surface may be directly observed during operation infrared radiation measurement may be used to measure temperature. Originally, Bowden [39] used this method to

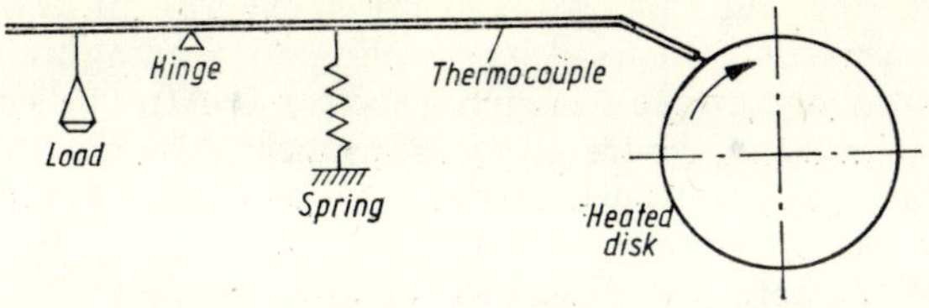

Figure 3.4 Trailing thermocouple.

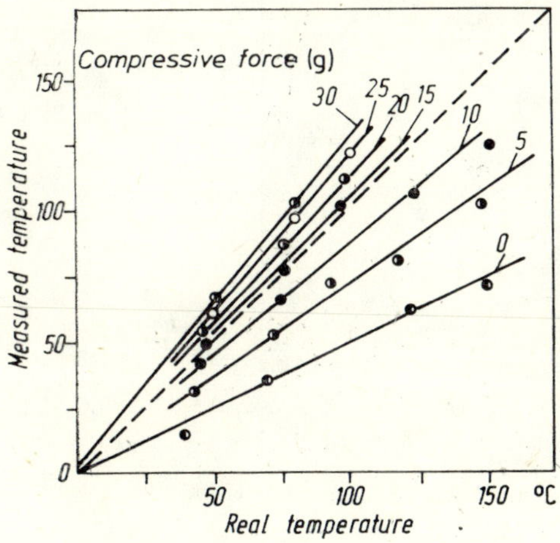

Figure 3.5 Dependence of contact thermocouple measured temperature upon real temperature for various compressive forces, without lubrication [57].

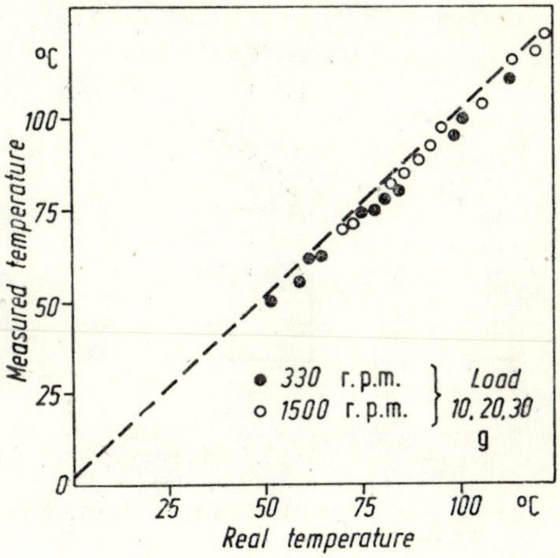

Figure 3.6 Dependence of contact thermocouple measured temperature upon real temperature for some compressive forces and speeds, whith lubrication [57].

measure the instantaneous temperature of a dry friction process between a glass surface and a metal pin. Under such conditions, owing to adsorption of infrared radiation by the glass, the technique gives good results only for high temperatures ($t > 500°C$).

To extend the use of the method to the measurement of lower temperatures, reduced adsorption of the infrared radiation is necessary. A simple solution is the substitution of the refraction process by reflection using metallic mirrors to reflect the radiation beam. To compensate for the spectral responsivity of the transducer and the adsorption of the beam during the path length, calibration of the system is necessary. This is achieved by means of a static heat source the temperature of which is measured by a thermocouple. Such a calibration involves a certain error because, although the radiating power of both surfaces may be chosen to be similar, these may not be identical. Any error due to different powers of radiation and changes in transducer characteristics with time may be reduced to a minimum , if reference to a standard is carried out during the period of actual temperature measurement and the calibration temperature is close to that of the surface temperature.

Figure 3.7 shows the basic diagram of a process unit where comparison between the measured value of the temperature and that of a standard surface temperature is carried out during the friction experiment. The friction couple consists of parts marked *1*, the fixed part having an opening to allow the infrared radiation of the moving surface to reach a mirror *2*. After reflection by mirror *2* the infrared radiation is concentrated on transducer *4* by mirror *3*. The temperature of the standard surface 5 measured by thermocouple *6* is directed onto the concave mirror *3* by turning mirror *2*. Use of a suitable transducer has extended the lower limit of temperature measurement down to 50°C.

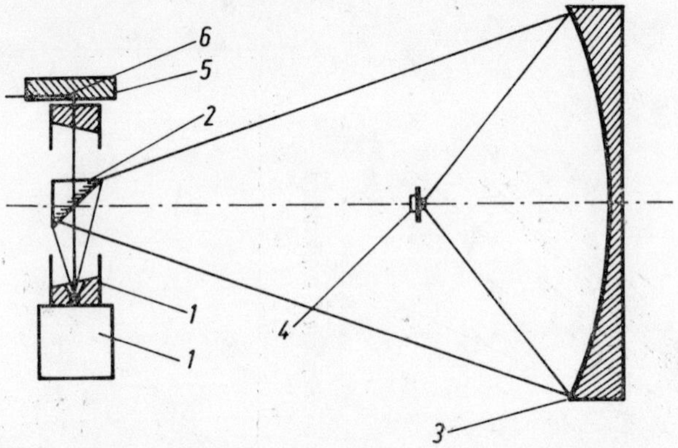

Figure 3.7 Measurement of surface temperature by infrared radiation [55]: *1*, couple; *2*, plane metal mirror; *3*, concave mirror; *4*, radiation transducer; *5*; standard surface; *6*, thermocouple.

Irreversible processes. With the exception of the dynamic thermocouple method the temperature measurement methods described require modifications to the friction couples which disturb the surface continuity and thus affect the value of the temperature measured. This drawback may be avoided if use is made of irreversible changes in the surface material, dependent upon temperature. Use of this principle provides a method of maximum temperature measurement. This is not particularly disadvantageous in practice as the maximum temperature is of interest.

One method of assessing the maximum temperature achieved by frictional contact is by the study of the metallographic structure of the superficial layer. For example exceeding the transition temperature causes irreversible structural changes which can be detected either directly or indirectly, for instance by hardness measurement.

A modern method, based on the same principle, utilizes the rate of kryptonate decomposition which is temperature dependent [51]. Radioactive krypton ^{85}Kr is introduced into the surface the temperatur eof which is going to be measured either by diffusion or bombardment by ions. The depth of penetration varies, according to the rated conditions, between 0.1 and 10 μm. A small depth of penetration affords a high degree of precision to the method.

By heating the surface to a specific temperature, an amount of the kryptonate present decomposes, degasing krypton. Cooling and reheating to temperatures below that reached during the first heating does not modify the existing quantity of kryptonate present in the surface. To calibrate the procedure the surface radioactivity is registered following a series of heatings to various temperatures (Figure 3.8).

To establish the surface temperature of the specimen after test, the surface radioactivity must be registered by progressive oven heatings. The maximum temperature value reached by the specimen surface during frictional contact is indicated when the measured radioactivity diminishes. The precision of the method

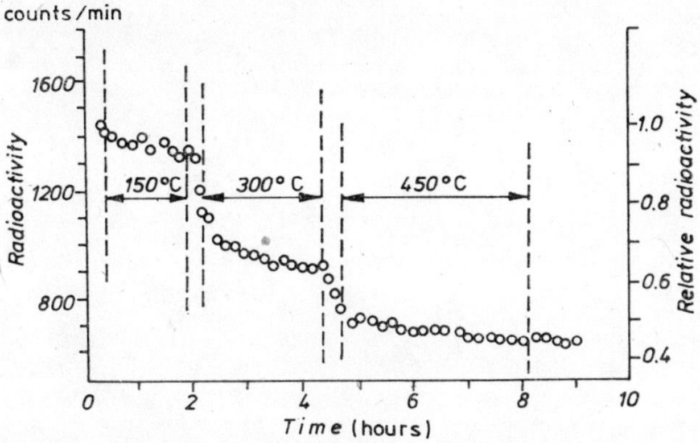

Figure 3.8 Radioactivity dependence upon the temperature, Kriptonate method.

depends upon the statistical error inherent in the measurement of the radioactivity and upon the selected temperature range. The precision generally obtained is 8°C for a 27°C temperature level and a measurement range of 0—600°C [51].

This method, unlike the technique dependent upon structural alterations, has the advantage of being applicable over a large range of temperature. Radioactivity measurement is simple as krypton 85 has a β active emission of 0.7 MeV.

As krypton is biologically inactive, this method does not require any special precautions being taken. Kryptonates have also been used to assess temperatures under difficult conditions such as the surface temperature of gas turbine blades or cams and tapets of internal combustion engines [204].

4 Thin layers

Friction between clean metallic surfaces is characterized by a high coefficient of friction and severe wear due to the specific properties of the surfaces such as high surface energy, reactivity and mutual solubility.

Clean surfaces also readily adsorb even traces of foreign substances such as gases, vapours and organic substances. The newly formed surface generally has a much lower coefficient of friction and wear than the clean surface.

Systematic studies of this behaviour have indicated that friction is reduced by the existence of a thin layer of material different from that of the contacting materials and which is strongly adherent to the solid surface. The almost general presence of a layer of foreign material on the surface of solid materials (perfectly clean surfaces are difficult to obtain and maintain) is related to the fundamental surface characteristic of uncompensated molecular (atomic) forces. This characteristic is responsible for the existence of surface properties such as surface energy and adsorption capacity.

The effect of thin layers on the frictional process is primarily due to the fact that displacement takes place between the layers and not between the metallic surfaces. It must be emphasized however that the thin layer structure provides the material forming this layer with new frictional characteristics. Most materials which in thin layer form have good frictional properties cannot be used as such due to their lack of cohesion. The presence of a sub-surface changes the characteristics of these materials so that the resulting layer-support sub-surface system allows tangential displacement by a smaller force with a reduction in wear even under heavy normal loads.

An important characteristic of thin surface layers is that their existence and frictional properties are independent of or only slightly dependent on the relative surface speed. These properties which exist even at low speeds or under static conditions aid hydrodynamic lubrication which is dependent upon surface displacement.

For thin layers of the type considered here the load carrying capacity is due to the forces of adhesion between the layer and the supporting sub-surface and the cohesive forces within the layer.

With thin layers, friction pairs may be obtained having frictional characteristics independent of the mechanical strength characteristics of the bulk material. Thus using thin layers the optimum operation of machine part regarding friction and mechanical strength can be obtained.

Thin layers used to improve the friction and wear characteristics of surfaces fall into several classes depending on the following criteria:

(a) technique of formation:

a_1 — layers formed spontaneously by the interaction between the metallic surface and the environment in which the friction pair operates. This layer may be renewed during operation;

a_2 — layers formed on the surfaces of the friction pair before they begin to operate. This layer is not renewed during operation.

(b) participation of the material of the friction pair:

b_1 — the surface material participates in the formation of the layer;

b_2 — the surface material does not participate in the formation of the layer.

(c) the mechanism of formation and the nature of the bond between the layer and the bulk material:

c_1 — adsorption/ Van der Waals bond;

c_2 — chemisorption/ chemical bond;

c_3 — reaction/ chemical bond;

c_4 — diffusion/ common crystalline lattice;

c_5 — deposition of material/mixed bond (e.g. adhesion + diffusion, possibly);

(d) thickness of the layer:

d_1 — monomolecular layer (thickness: one molecule);

d_2 — polymolecular layer (thickness: several molecular layers);

d_3 — massive layer (thickness of several tenths of a millimeter).

The particular types of thin layers may be classified according to the above criteria as follows:

E.g. Oxide layer formed on the metallic surfaces of the friction pair under normal atmospheric conditions

$$a_1\, b_1\, c_3\, (c_4)\, d_2.$$

The strength of the bond between the layer and the bulk material plays an important role for all types of thin layers. For massive deposited layers it is difficult to ensure an effective bond between the layer and the surface. In such cases, intermediate layers are used to improve the adhesion of the layer, for example, multiple metallic layers (Chapter 4.4).

The coefficient of thermal expansion is a factor of great significance for the endurance of massive layers differing in nature to that of the support sub-surface material. A considerable difference in the coefficient of thermal expansion between the material of the layers and the material of the friction pair will cause cracking of the layer due to temperature variation in operation.

From the point of view of surface coverage, the most commonly used layers are those formed spontaneously a_1 by the action of oxygen and lubricants (including additives) on the metallic surface.

Layers of type a_2 preformed by surface treatment except for those obtained by heat-treatment have a specific hence restricted application range. Table 4.1 lists several important techniques of surface treatment.

Table 4.1

Techniques of surface treatment (after [199])

Material added to surface	Surface chemistry altered	Surface microstructure altered
A. Welding — gas — arc — plasma B. Thermal spraying: — flame — arc — plasma — detonation C. Cladding: — brazing — explosive bonding D. Miscellaneous: — spark hardening — powder coating — organic finishes E. Electrodeposition: — electrolysis (aqueous) — electrophoresis — metalliding — ion treatment F. Vapour deposition: — physical — chemical G. Chemical deposition	H. Interstitial hardening: — carburizing — nitriding — carbonizing — cyaniding — sulphonitriding — boriding I. Diffusion treatment: — siliconizing — aluminizing — chromizing — sulphurizing J. Chemical treatment: — oxidation — phosphating	K. Mechanical treatment — peening — rolling — machining L. Thermal treatment M. Thermomechanical treatment — martensite work hardening — multiple processes
e.g. Plating followed by diffusion treatment		

The successful use of thin layers to improve the friction and wear characteristics of surfaces requires consideration of the following factors:

(a) materials are required which are capable of forming thin layers and their frictional characteristics determined;

(b) the techniques of layer formation should be investigated as to their effectiveness with the surface material used to obtain optimum service performance.

This chapter covers the following types of layers:

— homogeneous organic layers,

— preformed reaction layers (conversion layers),

— layers formed by solid lubricants with lamellar structure,

— soft metallic layers.

The effect of the oxide layer will be discussed in Chapter 5.2.

The subchapter "Homogeneous Organic Layers" deals with organic layers (hydrocarbons, fatty acids, alcohols, etc.) used in the absence of a fluid lubricant to allow examination of the basic frictional properties and structure of such organic layers. The results obtained, however, may have important practical applications when a fluid lubricant is present although the formation of the layer is also affected by the presence of lubricant molecules.

The subchapter "Preformed Reaction Layers" covers only those layers resulting from a previous surface treatment (sulphonitriding and phosphating). Layers formed by the reaction of extreme pressure or antiwear additives with metallic surfaces are not included.

The other two subchapters deal with the well-defined fields of lubricants with a lamellar structure and of soft metallic layers.

4.1 Homogeneous organic layers

The frictional behaviour of homogeneous organic layers, that is, layers consisting of a single type of molecule, is important in surface lubrication and has interested many authors (Langmuir, Bowden, Tabor, Zisman, Russell, etc. [39, 78, 92, 120, 112, 182, 152]).

These studies were partly carried out on monomolecular layers transferred from a water surface (Figure 4.1). Although this technique allows a thorough investigation of the deposited film as to its structure, number of layers and orientation of the molecules, the properties of layers thus obtained are considered to be different from those of layers obtained in practical applications.

This technique has now been replaced by adsorption from the melt. The surface on which the layer is to be deposited and which has previously been processed and cleaned, is immersed into the melt of the coating material. The excess material is removed by wiping with cotton-wool or filter paper. Although this technique, if incorrectly used, may introduce errors due to the uncontrollable thickness of the deposited layer, it is preferable to the earlier technique as it produces films similar to those obtained in practical applications.

Another technique which produces homogeneous monomolecular layers by adsorption from a solution has been described by Zisman [182]. Any polar or non-polar substance may be adsorbed if the solvent is less adsorbed and has a considerably higher surface tension than that of the resulting film. Prior to film deposition the surface of the solid material should be carefully cleaned to remove contaminated layers resulting from previous mechanical processings (turning, grinding, polishing), as surface contamination prevents the formation of continuous homogeneous layers.

For metallic surfaces the thickness of the oxide layer also plays an important role. Under normal working conditions the presence of an oxide layer cannot be avoided. The existence and thickness of this layer may be controlled by the surface processing technique. It is essential that the technique of surface machining prior to surface layer formation should be studied in order to ensure reproducible surfaces with well-defined frictional properties [94]. Surfaces roughness should be as low as possible considering the thinness of the surface layers.

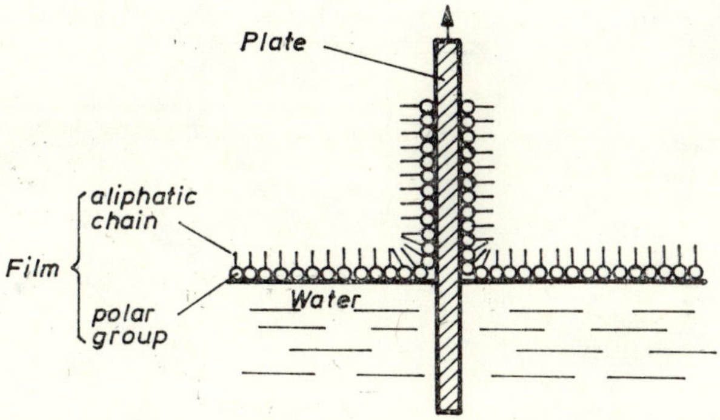

Figure 4.1 Transfer of an insoluble film.

The most common test configuration of a friction pair used for the study of thin layers is point or line contact. The test apparatus should be capable of low, variable speeds, for the measurement of the frictional and normal force and to ensure accurate control of the gaseous atmosphere. It should also allow easy mounting and dismounting of the friction pair. Figure 4.2 shows an apparatus used for the study of homogeneous thin layers [92, 93]. The friction pair consists of a bush, *1*, and a plate, *2*. The bush, which is mounted at the end of a shaft, *4*, is fixed by a nut, *3*. The plate support, *5*, is designed to allow rotation of the plate around an axle normal to the driving shaft so that line-contact always occurs between the bush and the plate. Thus, a uniform load distribution is achieved over the contact surface even if design or mounting defects exist. The plate support, *5*, which is embedded or hinged at one end has a cross profile with two sections having considerably different moments of inertia.

This arrangement allows simultaneous measurement of the load and frictional force. Although both the normal and tangential force cause bending deformations in both cross sections, the use of pairs of series-connected tensometric transducers allows separate measurement of the two stresses. Depending upon the desired measuring range, the moments of inertia of the two cross sections may be precisely selected. Load is applied externally through a pin, *6*. The complete apparatus is enclosed in a metallic case, *7*, and a cylindrical glass tube, *8*, so that the gaseous medium in which friction takes place may be controlled. To heat the friction pair, the clamp nut, *3*, is replaced by a metallic sleeve provided with an electric resistance (not represented in the figure).

With this apparatus a maximum temperature of 250°C can be obtained on the bush surface without the force measurements transducers being disturbed.

Measurement of the temperature of the contact surface is carried out by means of a miniature thermocouple with a stainless steel sheath mounted in the bush groove so that its junction is in contact with one of the bush lateral sides as

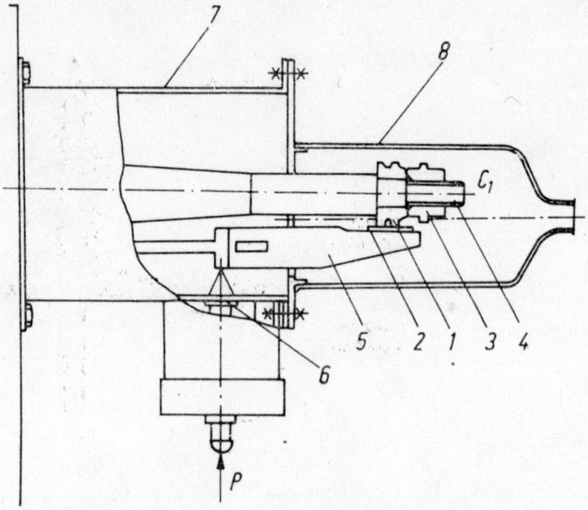

Figure 4.2 Disk on plate machine for the study of thin layers [93].

close as possible to the frictional surface of the bush. Location of the thermocouple in the bush groove ensures better heat transfer conditions than when located on the steel plate. The required contact pressure was determined by trial so that the measured values were stable when the bush was rotating. The calibra-

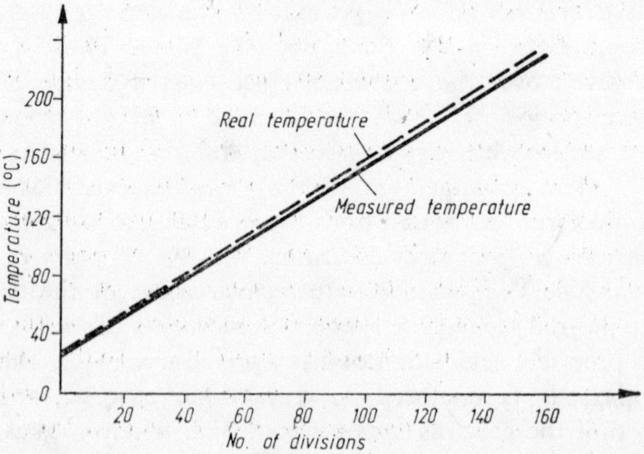

Figure 4.3 Calibration curve of temperature measured by contact thermocouple.

tion curve for the thermocouple used determined by means of a precision thermometer in an oil bath is shown in Figure 4.3. As the thermocouple does not measure the actual surface temperature in the frictional area a correlation between the actual and measured value required to be determined. For this purpose a phenomenon which is characteristic of the frictional contact of metallic layers, namely the discontinuity of the friction coefficient curve as a function of the temperature around the melting point of the metal was used. Using Woods' alloy (melting point = 68°C) and indium (melting point = 156°C) the calibration curve for the actual contact surface temperature could be plotted. A small difference between the mean actual temperature of the contact surface and the temperature measured by the thermocouple was noted.

4.1.1 **Effect of load and speed**

Using the apparatus described previously and a friction pair consisting of a bronze bush (6% tin) on a quenched high speed tool steel plate the variation of the coefficient of friction with load was established for some organic layers deposited by adsorption from the melt. The measurements were made at a sliding speed of $V = 0.4$ cm/s and a temperature of 25°C in a normal room atmosphere.

Figure 4.4 shows the significant effect of thin surface layers on the coefficient of friction and its independence of load. From values of 0.6 to 0.7, characteristic of clean surfaces, the coefficient of friction decreases to values from 0.13 to 0.03 with a thin surface layer. This effect on the coefficient of friction is accompanied by a considerable reduction in surface wear.

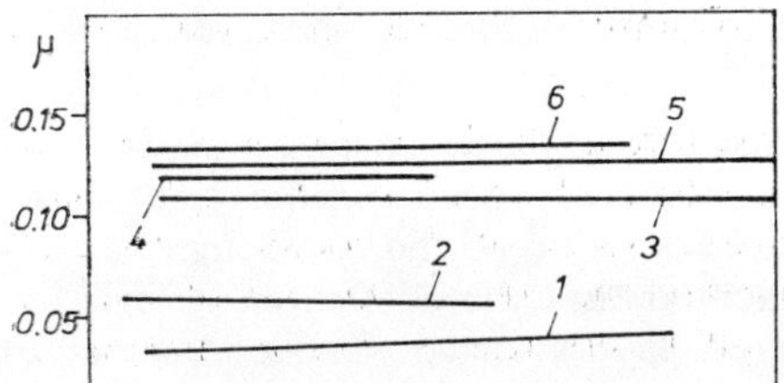

Figure 4.4 Variation of coefficient of friction with load for several substances deposited in thin layer form:
Sliding speed, V =0.4 cm/s; temperature, θ= 25°C: *1*, copper laurate; *2*, paraffin; *3*, MoS_2; *4*, cetyl alcohol; *5*, phosphorus additive, calcium base grease; *6*, dodecanoic acid.

Various types of layer deposition techniques were tested to obtain stable values for the coefficient of friction.

Thus, for a paraffin layer a considerable increase in the coefficient of friction was noted in one instance under a load of 30 daN. Using a lighter load the coefficient of friction did not return to its initial low value indicating that the protective layer had been destroyed.

After assessing various deposition techniques it was concluded that too thin a layer of paraffin was responsible for this result. Using a satisfactory film formation technique, constant values up to a load of 50 daN were obtained. The low value of the coefficient of friction is an interesting characteristic of the paraffin layer.

Lauric acid however, although it possesses an active group, has a higher coefficient of friction. If the substance (lauric acid in this instance) has insufficient time to react owing to the layer having been deposited on a freshly prepared fric-

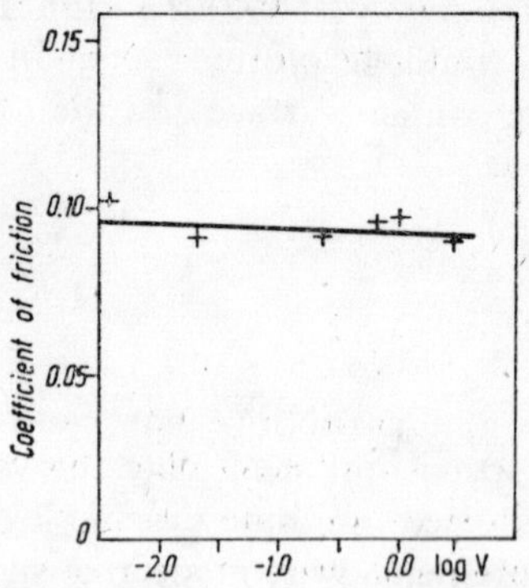

Figure 4.5 Coefficient of friction at low sliding speeds. Steel surfaces lubricated with pelargonic acid [39], (speed in cm/s).

tion pair, the coefficient of friction is affected only by the melting point of the respective compound.

The formation of pure layers of organic acids in normal atmosphere (relative air humidity of approximately 40%) has been found to be difficult. Unless special precautions are taken a reaction layer in addition to the chemisorbed layer is formed. Formation of the reaction layer is aided by an increase in temperature and the presence of thicker oxide layers on the surfaces. With freshly prepared surfaces in an atmosphere of purified argon constant values for the coefficient of friction were obtained even for fatty acids.

Determination of the coefficient of friction at high speeds was difficult because severe wear of the layer occurs at increased speed. Under light loads of only 5 daN measurement of the instantaneous value of the coefficient of friction at the moment of load application indicated that the coefficient of friction is independent of speed up to 60 cm/s for all layers tested. The coefficient of friction is also independent of speed for very low speeds. Figure 4.5 shows the variation of the coefficient of friction with speed for sliding speeds from 0.01 cm/s to 1 cm/s for a layer of pelargonic acid.

4.1.2 **Effect of the nature of the layer**

The effect of the chain length of the carbon atoms of acids, alcohols and aliphatic amines is shown in Figures 4.6, 4.7, 4.8. Besides the coefficient of friction determined at a speed of 0.01 cm/s, the value of the contact angle α of the layer with respect to methylene iodide is given. The friction pair used was a stainless steel sphere of $\Phi = 12$ mm diameter and a glass plate.

In each instance a minimum number of carbon atoms exists (14 or 18) after which the coefficient of friction and the contact angle become constant. This relationship, taking into account the fact that the surfaces of the pair were non-reactive, indicates that a minimum coefficient of friction is obtained when the layer reaches a maximum state of condensation. Such a state, which is possibly

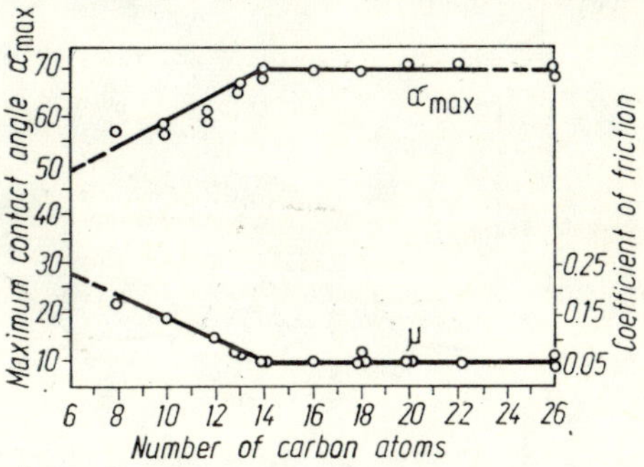

Figure 4.6 Coefficient of friction and contact angle versus number of carbon atoms for aliphatic acids [182].

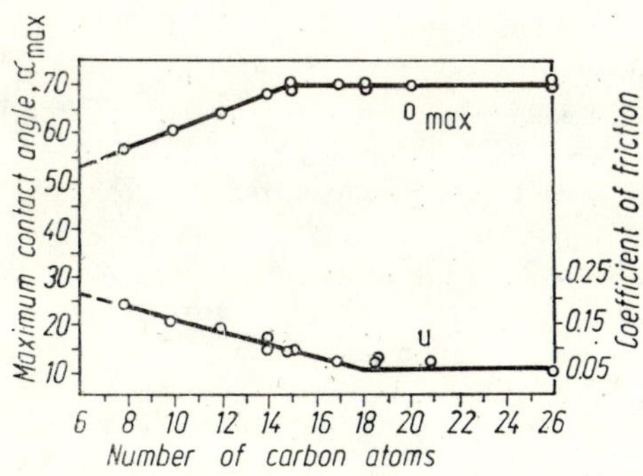

Figure 4.7 Coefficient of friction and contact angle versus number of carbon atoms for aliphatic alcohols [182].

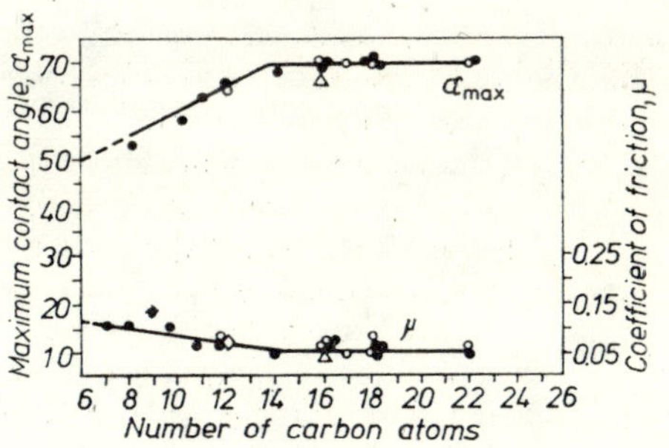

Figure 4.8 Coefficient of friction and contact angle versus number of carbon atoms for primary amines [182].

accompanied by a favourable orientation, is achieved for polar molecules at a shorter chain length than for non-polar molecules. In other words, the polarity of the molecule compensates for chain length.

4.1.3 Effect of temperature and the materials of the friction pair

Using the apparatus shown in Figure 4.2 the effect of a temperature rise on the properties of the layers was established.

Figure 4.9 shows the effect of temperature on a paraffin layer. Around the melting point the coefficient of friction increases from 0.06 to 0.25. This phenomenon is reversible, demonstrating the relationship between the increase in the coefficient of friction and the melting point of the material of the layer. Upon further increase in temperature a linear rise in the coefficient of friction is observed. Starting from a certain temperature, a rapid and irregular growth in the coefficient of friction occurs, due to either evaporation of the layer or deterioration of the surface of the bronze bush by oxidation.

Although the melting point of the layer material is well-defined the coefficient of friction does not however rise sharply when the layer reaches its melting

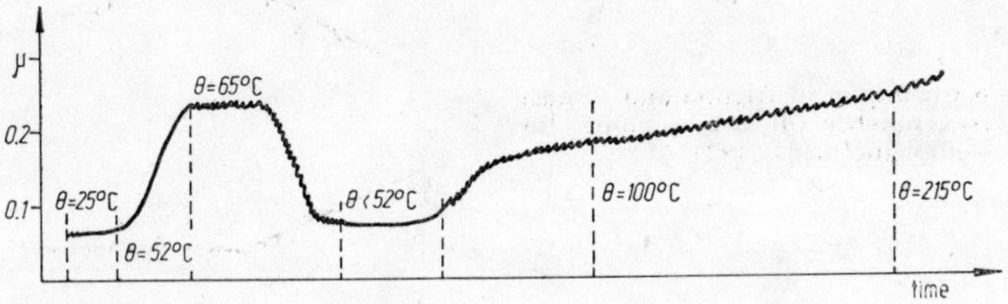

Figure 4.9 Effect of temperature θ on the coefficient of friction of a paraffin layer.

point probably because transition to the liquid state does not occur simultaneously over the whole surface. The discrepancy between the time when the coefficient of friction begins to increase and the time the melting point is reached is accounted for by the time discrepancy between the actual and the measured temperatures which depends on the rate of heating.

With vaseline oil, a liquid under ordinary temperature conditions, instead of paraffin, a continual increase in the coefficient of friction was obtained, similar to that for a paraffin in the liquid state. This linear dependence may be accounted for by increase in the degree of disorder in the layer. It should be noted that the coefficient of friction, although high in the presence of a non-polar hydrocarbon even in the molten state, is lower than the coefficient of friction for clean metallic surfaces (0.25 to 0.3 as compared with 0.6 to 0.7).

If a polar, non-reactive substance is used for surface layer formation increase in the coefficient of friction at the melting point is not so pronounced. Thus, with lauric alcohol, the coefficient of friction only increases from 0.09 to 0.16 (Figure 4.10).

With lauric acid, at temperatures below the melting point, the coefficient of friction is not stable, varying between 0.1 and 0.15. Exceeding the melting point results in reduced fluctuation by elimination of the low values. A further increase in temperature results in a slow linear increase in the coefficient of friction.

Investigation of the effect of temperature under these conditions is difficult to carry out as continued repeated heating of the friction pair causes formation of a reaction layer which gradually changes the frictional behaviour. As it is not probable that all the acid reacts simultaneously it may be concluded that if two active substances are present on the friction surface, the resulting coefficient of friction will be determined by the substance having the lower coefficient of friction. The absence of a transition temperature in the variation of the friction coefficient of lauric acid with temperature for a bronze on steel pair may be accounted for by the formation of iron laurate the melting point of which is higher than its decomposition point. With a bronze on bronze friction pair, in the absence of iron laurate, the transition temperature is lying approximately at 110°C, the melting point of copper laurate (Figure 4.11).

Investigations of the behaviour of organic layers are generally carried out over a temperature range from room temperature to 200°C. To extend the temperature range from −250°C to +250°C special devices such as that illustrated in Figure 4.12 have been designed [152].

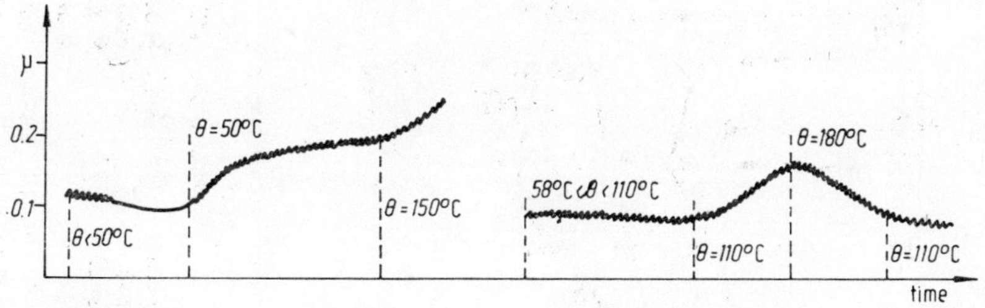

Figure 4.10 Effect of temperature on the coefficient of friction of a layer of lauric alcohol.

Figure 4.11 Coefficient of friction versus temperature θ for the lauric acid layer and bronze surfaces.

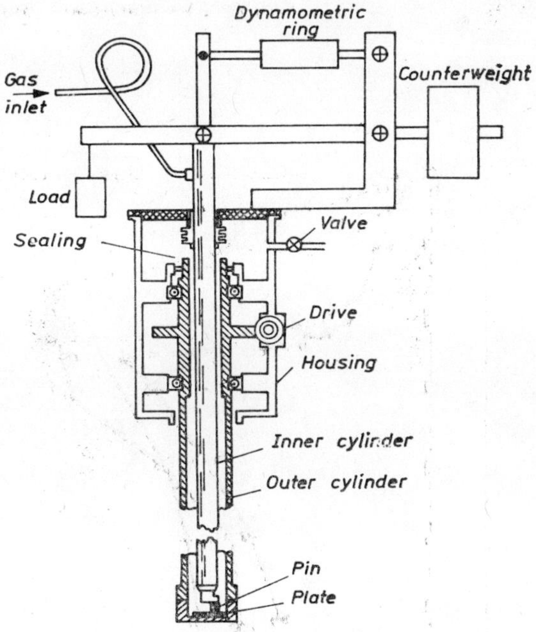

Figure 4.12 Apparatus for the study of thin layers behaviour over a large temperature range [152].

In the apparatus illustrated in Figure 4.12 the friction pair mounted at the lower end of the shaft consists of a spherical pin pressed against a rotating plate. This permits cooling of the friction pair by immersion of the lower part of the apparatus into a cooling fluid. A gas flow may be introduced between the outer sleeve and the driving shaft to control the atmosphere in the area of frictional contact.

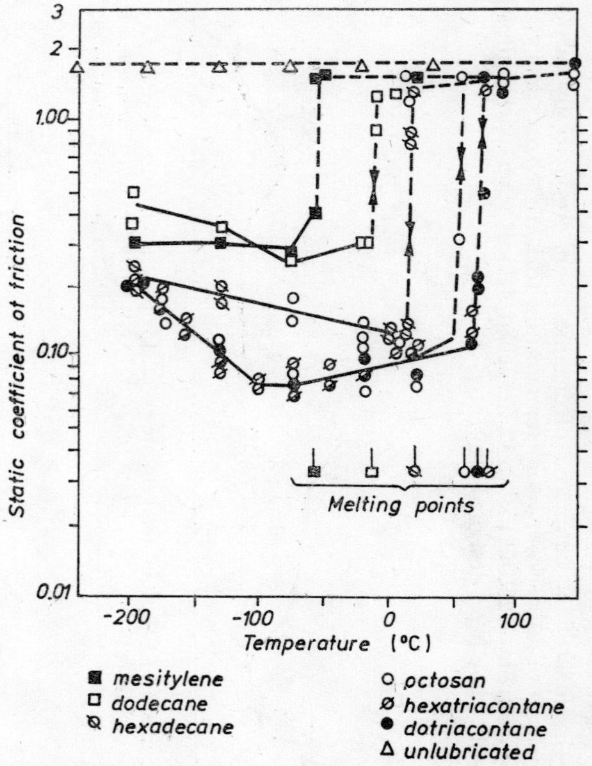

Figure 4.13 Coefficient of friction versus temperature for several hydrocarbons in helium atmosphere [152]. Friction pair : copper on copper.

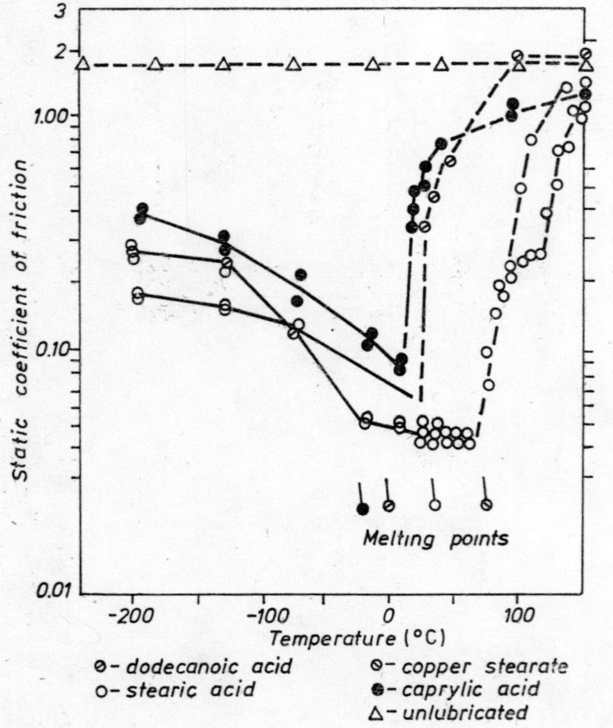

Figure 4.14 Coefficient of friction versus temperature for several aliphatic acids in helium atmosphere [152]. Friction pair: copper on copper.

With this apparatus the effect of temperature on the coefficient of friction was determined for several organic layers over a temperature range of $-190°C$ to $+200°C$ (Figures 4.13 and 4.14), under a load of 400 g and at a sliding speed of 0.01 cm/s. Both parts of the pair were made of copper.

As previously found the existence of a transition point was observed in the coefficient of friction corresponding to the melting point of the compound. The carbon atom chain length has also a beneficial effect. An interesting characteristic is the continuous decrease in the coefficient of friction up to the transition point. If the layer, although reactive, cannot react with the surfaces of the friction pair due to the prevailing conditions, the transition temperature coincides with the melting point of the layer (Figure 4.14).

4.1.4 Layer durability

In addition to a low coefficient of friction, the deposited layers should have a long wear life which depends on the properties and thickness of the layer. Measurements carried out on the apparatus shown in Figure 4.2 confirmed the importance of the deposition technique on the wear properties of the layer. The deposition technique should be selected to ensure satisfactory removal of any excess coating substance. The thickness of the resulting layer should be function of the adhesion properties of the active coating substance. As the result of several tests the following deposition technique was found to be satisfactory:

— formation of the layer by immersion of the surfaces of the friction pair into the molten active coating substance;

— 2 min. running under the long time test load; this procedure removes the excess active coating substance from the loaded area;

— rubbing against filter paper under a light load for 1 min.; this procedure removes the excess of active coating substance in the vicinity of the loaded area. It also acts like a mild wear process in reducing the layer thickness down to a value determined by the layer adhesion to the surfaces of the friction pair;

— wiping off of the excess coating substance from the plate.

Such a deposition technique aids selection of the active coating substances as the wear-resistant layers are also resistant to mild abrasion.

Figure 4.15 shows results obtained with a bronze on quenched high speed tool steel friction pair under a load of 5 daN, at a sliding speed of 0.4 cm/s for paraffin, MoS_2, lauric acid and copper laurate.

The results obtained using the above deposition technique are denoted by *1* while results obtained in the presence of a certain amount of excess substance are denoted by *2*. This excess was obtained by eliminating the penultimate phase of the above technique. As expected, paraffin had the lowest resistance, followed by lauric acid and copper laurate. With excess, the resistance of the copper laurate exceeded 14 hours.

The carbon chain length also affects layer wear life.

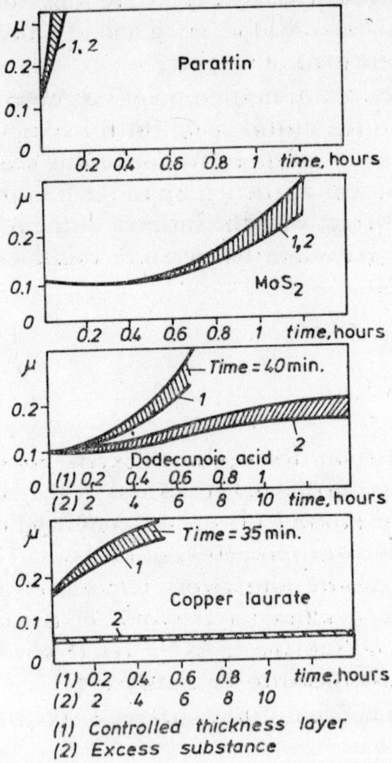

Figure 4.15 Wear life of several thin layers.

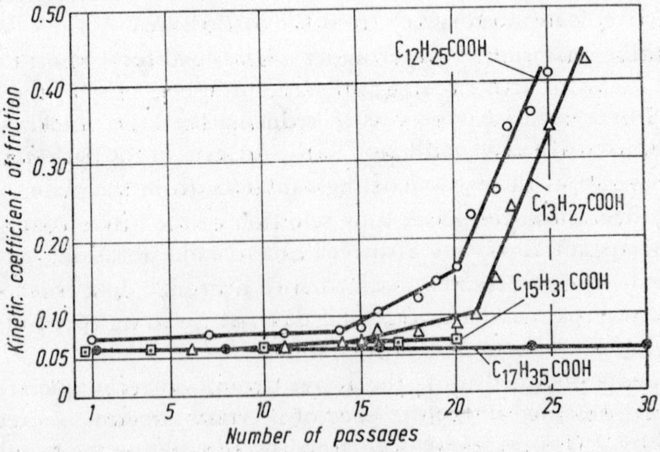

Figure 4.16 Resistance to wear versus number of carbon atoms in the molecule of a fatty acid [182].

Figure 4.16 shows variation of the coefficient of friction with the number of passages for several aliphatic acids; layer resistance depends on the carbon atoms chain length.

4.1.5 Structure and mechanism of action of adsorbed thin layers

Initial data concerning the structure of thin layers with special frictional properties were obtained when insoluble films were investigated. It is known that many substances which are insoluble in water spread over the surface of water to form a film which in most cases is monomolecular. By examining the relation $\pi = f(s)$, i.e. the film pressure as a function of the surface occupied by a molecule, it was concluded that several types of layers exist. The characteristics of these layers exhibit a certain resemblance with one of the three states of aggregation (solid, liquid and gaseous). It was found that for fatty acids in the aliphatic series the surface occupied by a molecule in a solid film is of 20.5 Å, irrespective of the number of carbon atoms in the molecule, provided this number is greater than 16. Assuming that the space occupied by a molecule is cylindrical in shape and has a constant base area, any expansion of the molecule volume may occur only by an increase in height.

Thus, it was concluded that filiform molecules should be normal to the surface. As these films exhibit a hydrophilic face towards the water and a hydrophobic face in the opposite direction (as seen in Figure 4.1), it follows that the constituent molecules have their polar group oriented towards the water.

Films of this type, namely with parallel molecules oriented almost normal to the surface are also formed by salts, amines, alcohols, etc. These films have unique frictional properties. Such a monomolecular film of barium stearate deposited on a clean surface was reported to have a coefficient of friction of only 0.05 to 0.06.

Structural studies carried out on such films by means of X-ray and electron diffraction have confirmed previous hypotheses. As the shape of the molecules is known the structure of these films can be represented more accurately.

The reaction of a fatty acid with a metallic surface produces a polymolecular soap film. The structure of such a film was established by examining the X-ray diffraction pattern on a mercury surface (Figure 4.17). The layer which forms is polymolecular and the surfaces with the same nature are conjugated. The degree of order increases from outside towards the metal surface. The greater the number of layers, the closer the structure of the layer to the structure of a crystal. The great resemblance between the above type films and a crystalline solid material follows also from the fact that the surface occupied by a molecule in a crystal plane is 18.5 Å^2 as compared with the surface occupied by a molecule in a solid film of 20.5 Å^2.

If the layer is formed by adsorption from solution, its structure depends on solvent concentration. Thus, for a concentration of only 0.001% of a fatty acid in normal-hexane after a 10 min. contact with the solution, the metallic surface shows no diffraction pattern. For a concentration of 0.01% the image obtained is diffuse while a clear diffraction pattern is obtained only at a concentration of 0.05%. Obviously, the solvent molecules, interposing among the adsorbed molecules, prevent a compact arrangement of the latter. As Chapter 6.1 will show, the effect of disturbance of the solvent (the lubricant in the case of additives) is also found with wear under boundary conditions.

Temperature is another factor which affects the ordinate structure of the layer. Near the melting point of the substance from which the layer was generated, the diffraction pattern becomes diffused or vanishes. This observation confirms the hypothesis that the low coefficient of friction of these layers is due to their structure which is characterized by a high density of the substance in the layer with poor adhesion between the layers. This structure, termed a lamellar structure

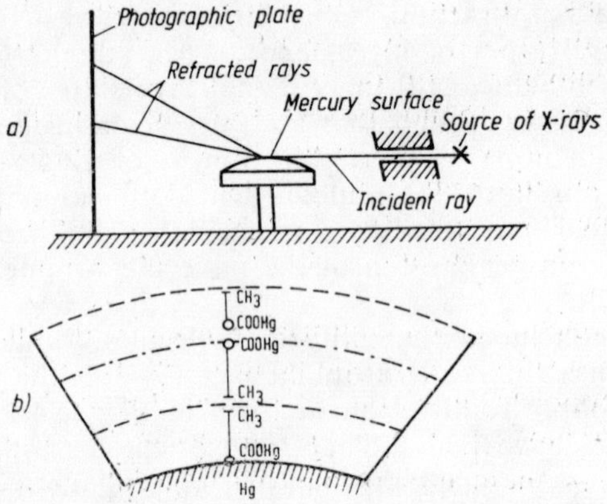

Figure 4.17 Use of X-rays for determining the structure of thin layers:
(*a*) apparatus; (*b*) structure of soap layer on mercury surface

also occurs in solid lubricants which have good frictional properties. As distinguished from these materials which preserve their lamellar structure independently of the presence of a surface, the formation of solid-type films is determined by the simultaneous existence of a surface and the filamentary form of the molecule.

In addition to the existence of a polar group, the formation of a high density adsorbed layer of organic compounds requires the remaining part of the molecule to be filiform. These conditions are best obeyed by the aliphatic compounds. The importance of the length of the chain of the carbon atoms lies in the fact that the energy of cohesion between adjoining methylene groups $-CH_2-$ contributes to the total energy of adsorption. If the polar group is sufficiently small in volume not to disturb the chains of carbon atoms then the total energy of cohesion is written as

$$U_c = u \cdot N, \tag{4.1}$$

where u is the energy of cohesion for a methylene group and N is the total number of carbon atoms.

The total energy of cohesion is obtained by summation of the energy of cohesion U_c and the energy of adhesion U_0 of the polar group to the surface.

$$U = U_0 + uN. \tag{4.2}$$

Thus the effect of the chain of the carbon atoms in a homologous series and the existence of a maximum value for N which is required to achieve a maximum density film ($N = 14$ for fatty acids) may be explained. The branching of the molecule or the introduction of other polar groups will cause a swelling of the film or will even prevent its formation. Account should be taken of the fact that, although unordinate, the molecules which are adsorbed on solid surfaces will contribute to the separation of these surfaces and reduce the friction force, without having the same effect as solid films.

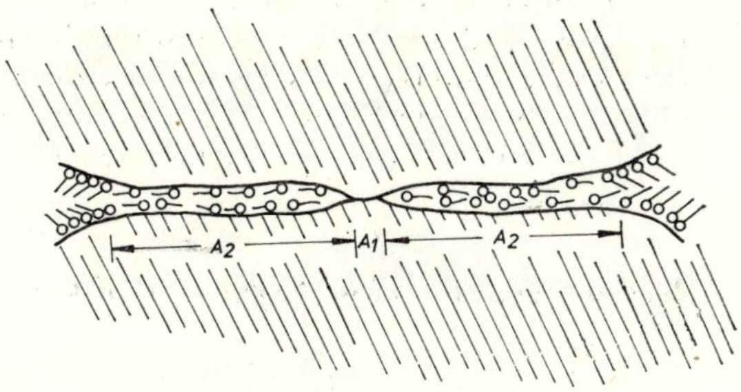

Figure 4.18 Boundary lubrication process: A_1, surface of contact between solids; A_2, surface of contact between compressed films.

To explain the effect of adsorbed layers on friction the hypothesis of the complete separation of the solid surfaces by the adsorbed layer was initially proposed; friction would take place between the outer surfaces of the films. A more complete examination of the behaviour of thin layers during frictional contact has indicated that this phenomenon is more complex. Direct contact between metallic surfaces through the layer exists under any load as established experimentally by means of radioactive tracers of metal transfer between surfaces for a single passage. The most important result of this investigation lies in the conclusion that the considerable reduction in transfer with lubrication by adsorbed thin layers is principally due to the size of the fragments and not to the number of fragments [39]. A new model for contact under boundary conditions is proposed based on the above considerations (Figure 4.18).

The friction force in this instance is given by the relation

$$F = A\,[\alpha\mu_m + (1 - \alpha)\,\mu_f], \tag{4.3}$$

where

F is friction force,
A — area of contact,
α — fraction of surface in metallic contact,
μ_m, μ_f — coefficients of friction for the metal-metal and film-film contacts.

For a film adsorbed on a metallic surface, α should be of the order of 10^{-4} to take into account the low metal transfer. It should, however, be noted that the value of the electrical contact resistance for surfaces covered with a monomolecular layer is about the same as for clean surfaces. If it is considered that the contact resistance is a quantity which is directly determined by the area of contact, a contradiction follows which is avoided only if the electrical resistance of the monomolecular layers is assumed to be, except for several particular instances, much smaller than that of the bulk material of the layer.

Considering the above, the contact between two surfaces covered with a monomolecular layer can be described as follows: as the pressure on a (spherical) asperity is non-uniform, metal-metal contact occurs in the high pressure area and the load is taken over by the film in the areas where the pressure is decreasing. Thus, the number of contacts for a given apparent area of contact remains about the same, whereas the metal-metal contact surface is considerably reduced.

This interpretation emphasizes the importance of the lateral cohesion between the molecules of the layer. Cohesion between the molecules is affected by temperature and the presence of some foreign molecules interposed among the molecules of the layer. The factor of cohesion between the molecules allows distinctions to be made between various substances used to form layers, as the process of adsorption is a phenomenon generally valid for any surface and fluid.

4.2 Preformed reaction layers (Conversion layers)

Layers formed by reaction between a metallic surface and a gaseous, fluid or in solution reactant, are termed conversion layers. An additional condition for such layers is to have a low solubility of the reaction product in the material of the surface and a high concentration on the surface. These additional conditions exclude from the group of conversion layers those layers obtained by carbonizing, although the mechanism of formation of such layers has several aspects in common with the formation of conversion layers. Some of the most effective conversion layers with regard to frictional contact are layers obtained by sulphonitriding or phosphating [178].

4.2.1 Sulphonitriding

The technique of sulphonitriding of ferrous materials represents a modification of the technique of nitriding in a cyanide bath by adding sodium sulphite. The following reactions take place at the working temperature (570°C) of the bath:

$$3NaCN + Na_2SO_3 = Na_2S + 3NaCNO, \tag{4.4}$$

$$Na_2S + Fe \rightarrow FeS. \tag{4.5}$$

Sodium sulphite acts also as a catalyst in the formation of cyanate by direct oxidation of cyanide

$$2NaCN + O_2 = 2NaCNO. \tag{4.6}$$

The cyanate formed by these reactions is the azote source

$$4NaCNO = Na_2CO_3 + 2N + CO + 2NaCN. \tag{4.7}$$

Below 450°C the cyanate bath provides a solution of interstitial azote, possibly saturated, on the surface. By annealing at 300°C, hard Fe_3N may precipitate in an acicular form. For medium hard steels it was found that the thermochemical treatment which results only in saturation with interstitial azote has no other effect than embrittlement of the surface subjected to frictional contact. Under the action of the friction force ferrite grains tend to displace in the direction of the nitride crystals. It appears that the only beneficial effect of treatment in a

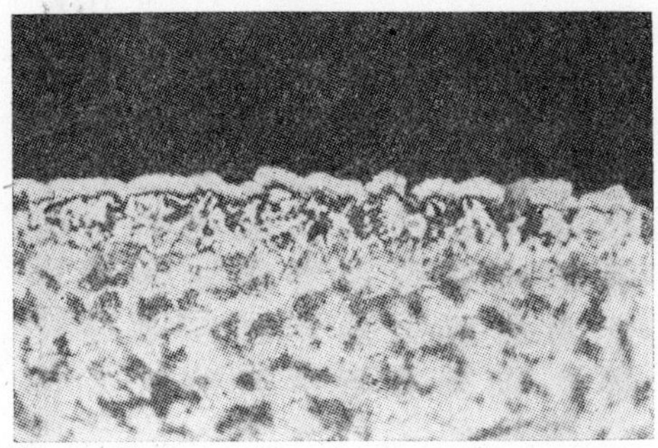

Figure 4.19 Structure of layer obtained by sulphonitriding.

syanate bath at temperatures below 450°C is to increase the surface fatigue ctrength [192].

At temperatures over 500°C cyanate bath treatment leads to the formation of a compact layer with a hardness of 900 HV to 1100 HV. The layer thickness is a function of the immersion time.

By the use of a sulphonitriding bath which contains sodium sulphide and sulphite in addition to cyanate, the resulting layer has two parts: an outer porous region and a compact region in the immediate sub-surface area (Figure 4.19). The porous layer of high hardness has a thickness of 8 to 40 microns depending upon the bath temperature and the immersion time. The beneficial effect produced by this layer during the frictional process is due to its porous structure which allows better retention of the lubricant. The compact region of the layer allows the desired surface finish to be achieved during running-in by plastic deformation only without wear of the surface.

More recent investigations have shown that mixed porous-compact layers can also be obtained by the use of cyanide-cyanate baths [124]. The presence of the compounds containing excess sulphur may be detrimental owing to the formation of a deposit of FeS which is difficult to remove. This type of baths may be divided into 3 groups depending on sulphur content:

(a) Baths with a low sulphur content. Due to the low corrosive nature of these baths the resulting layer is compact and thin (5—12.5 μm); specimen dimensional stability is not affected.

(b) Baths with a medium sulphur content. A relatively thick mixed porous-compact layer is formed. No important changes in specimen size occur. A black layer of adherent FeS is formed.

(c) Baths with a high sulphur content. Because of the high sulphur content, this bath is corrosive. A thin mixed porous-compact layer is formed; specimen dimensional stability is greatly affected. Optimum results are obtained by control of the ratio of sulphur to cyanide content of the bath (Figure 4.20). The distribution of elements in the layer is shown in Figure 4.21. Although the active elements

N and S are mainly concentrated in the layer, some are also found in the metallic surface.

Sulphonitrided surfaces have an increased wear life under both dry and lubricated conditions of frictional contact. Figure 4.22 plots the results obtained

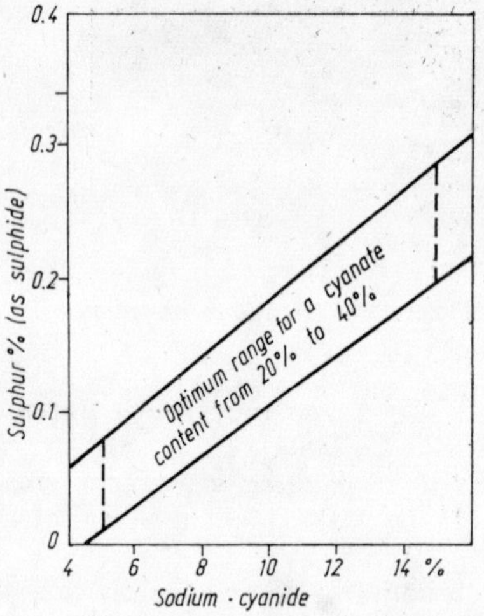

Figure 4.20 Optimum sulphur and cyanide contents for the sulphonitriding bath [124].

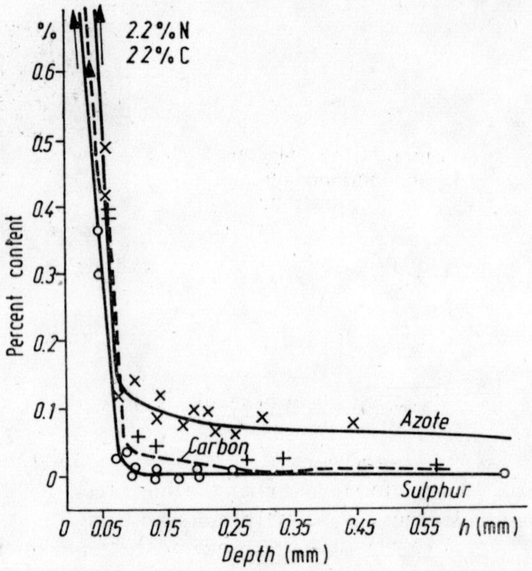

Figure 4.21 Concentration of N, C and S versus depth of a layer obtained by sulphonitriding [179].

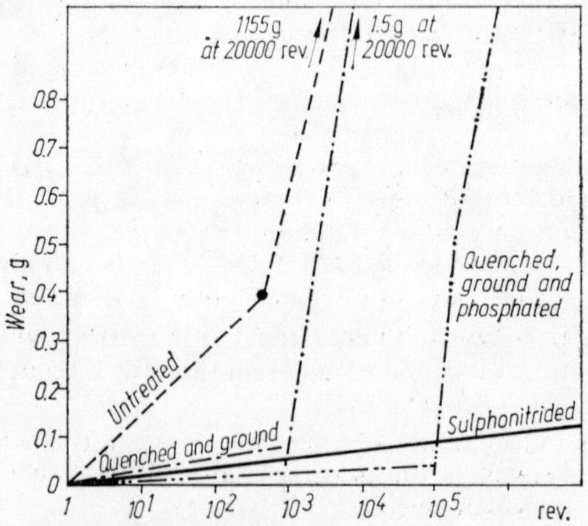

Figure 4.22 Wear life of several conversion layers [124].

on an Amsler machine with an additive-free lubricant. The layer is reported to be effective even after 10^6 rotations. For comparative purposes, Figure 4.22 also gives data obtained by other techniques of surface treatment.

The effectiveness regarding frictional behaviour of the layer obtained by sulphonitriding is due to its high hardness, porosity of the surface area and the presence of sulphur compounds having anti-seizure properties. An advantage of this technique compared with other thermochemical treatments is that it is directly applicable to finite parts. The dimensional changes of the part are negligible so that finishing after the treatment is unnecessary.

Sulphonitriding has produced good results especially with frictional pairs where mixed or boundary friction are predominant: for example cylinder sleeves for Diesel engines and spark-ignition, cam-tappet mechanisms, machine-tools guides and mechanisms with non-steady lubrication.

An improvement to the classical technique of sulphonitriding is effected by the additional use of an electrolysis process together with the thermochemical process which allows a reduction in the working temperature down to 250°C [192].

4.2.2 **Phosphating**

By immersion into an aqueous solution of a dihydrogen phosphate the surface of a metallic part becomes covered with an adherent layer of phosphate. Depending on the bath composition the resulting layer is either crystalline or amorphous.

The formation of a crystalline layer of phosphate occurs with acid solutions ($pH = 2$) of dihydrogen phosphates. Zinc and manganese phosphates are mainly used. Due to the acid medium and the double-exchange reaction, the solution also contains the ions of the metal to be treated. Consequently, the deposited phosphate layer is a mixed phosphate or a mixture of phosphates. Thus, during phosphating of a steel surface in an acid solution of zinc phosphate the reactions which take place are:

$$2H^{+} + Fe \rightarrow Fe^{++} + H_2, \tag{4.8}$$

$$Zn\,(H_2\,PO_4)_2 \leftrightarrows Zn^{++} + 2\,H_2PO_4^{-}, \tag{4.9}$$

$$3Fe^{++} + 3Zn^{++} + 4PO_4^{---} \rightarrow Zn_3(PO_4)_2 + Fe_3(PO_4)_2, \tag{4.10}$$

$$Fe^{++} + 2H_2\,PO_4^{-} \rightarrow Fe(H_2PO_4)_2. \tag{4.11}$$

At a temperature close to its boiling point the acid phosphate bath produces a sufficiently thick layer in about half an hour. To accelerate this reaction an oxidizer is used to depolarize the metallic surface. These accelerators reduce the bath temperature to 40 $\div$ 70°C and reduce the immersion time to several minutes.

The amorphous phosphate layer is obtained with neutral solutions of alkaline or ammonium phosphates and an oxidizer

$$4Fe + 2PO_4^{---} + 3O_2 + 6H^{+} \rightarrow 2FePO_4 + Fe_2O_3 + 3H_2O. \tag{4.12}$$

The characteristics of the phosphate layer, such as grain size, thickness and porosity greatly depend on the surface "history". Thus, previous surface cleaning with alkaline hydroxides results in a coarse layer structure whereas degreasing with organic solvents leads to a fine structure. Mechanical processing before the layer is deposited also affects its structure. These effects may be accounted for if consideration is given to the fact that the phosphating process is a topochemical reaction. An effective method of controlling the formation of the phosphate layer is to form some metallic centres to initiate the reaction by immersing the part into a dilute solution containing ions of an electropositive metal with respect to iron. The particles of deposited metal nucleate the phosphate crystals.

The structure and composition of layers deposited by phosphating have been studied by X-ray and electron diffraction. When steel surfaces were phosphated in a bath of zinc phosphate, the compounds $Zn_3(PO_4)_2 \cdot 4\,H_2O$ crystallized in the orthorhombic system and $Zn_2Fe(PO_4)_2 \cdot 4H_2O$ crystallized in the monoclinic system were identified. With a bath containing an alkaline phosphate and phosphoric acid, the layer consists of $Fe_3(PO_4)_2 \cdot 8\,H_2O$. An important conclusion of the structural studies was the epitaxial development of the phosphate crystals with respect to the crystalline structure of the metal. This explains the strong adhesion of the layers formed by phosphating to the support metal.

Phosphating like sulphonitriding has a significant effect on wear under mixed or boundary lubrication conditions (Figure 4.22). The effect of phosphating is more important as far as resistance to scuffing is concerned. Thus, for some hypoid gears phosphating increases the scuffing load by 1.4 times [111, 179]. This is important particularly during the running-in of gears. As the phosphating process is carried out at temperatures below 100°C, any alloy may be treated without involving structural changes.

4.3 Materials with lamellar structure

There are several substances which, due to the arrangement of the atoms in the crystalline lattice slip easily in a certain direction and thus the crystal may be compared with a package of lamellae (graphite, mica, boron nitride, molybdenum disulphide, etc.). Such substances, deposited on the surface of a friction pair, after a certain operating time form films with a preferred orientation in which the planes with the maximum atom density are parallel to the surface. These layers which have a strong internal cohesion prevent direct contact between the surfaces with little resistance to tangential displacement.

Some of the more frequently used lamellar substances namely: graphite, molybdenum disulphide and some substances related with the latter having special frictional properties will be discussed.

4.3.1 Graphite

Graphite was one of the first substances with a lamellar structure to be used as a solid lubricant. Its ideal crystalline structure shown in Figure 4.23 consists of parallel planes of carbon atoms arranged in a hexagonal lattice. The distance between the carbon atoms in these planes is 1.42 Å, whereas the distance between the planes is 3.40 Å. Because of the low binding energy between the atoms in adjoining planes with respect to the binding energy of the atoms in the same plane, graphite has a distinctive anisotropy of mechanical properties. Under the action of a relatively small force displacement of the planes with a high density of carbon atoms occurs. It has been shown that the presence of dislocations in the crystal

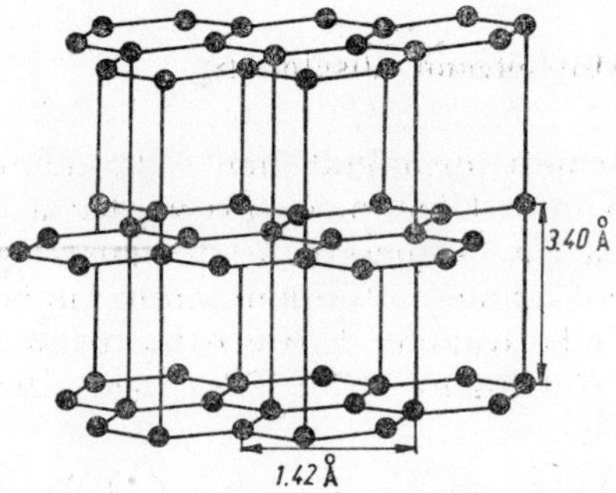

Figure 4.23 Structure of graphite.

lattice contributes to the free mobility between the extreme planes of a microcrystal. Consequently, in order to be effective for the frictional process, the graphite layer should have microcrystals oriented with the datum planes of the carbon atoms approximately parallel to the direction of motion. From the interpretation of electron diffraction patterns it was concluded that the graphite film formed on frictional surfaces has an ordinate structure and that the datum planes of the carbon atoms of the crystallites make an angle of 5° ÷ 10° with the support surface. In this state, the coefficient of friction has a minimum value of 0.15 to 0.18. By changing the direction of motion, the coefficient of friction increases but quickly returns to the low value.

A feature of graphite is the variation of its coefficient of friction with the gaseous composition of the environment. By degasing in vacuum at approximately 110°C the coefficient of friction increases to about 0.45. If small quantities of water vapour, oxygen or hydrocarbons are present, the coefficient of friction is reduced to the value in air (Figure 4.24) [150, 155]. As the quantity of water

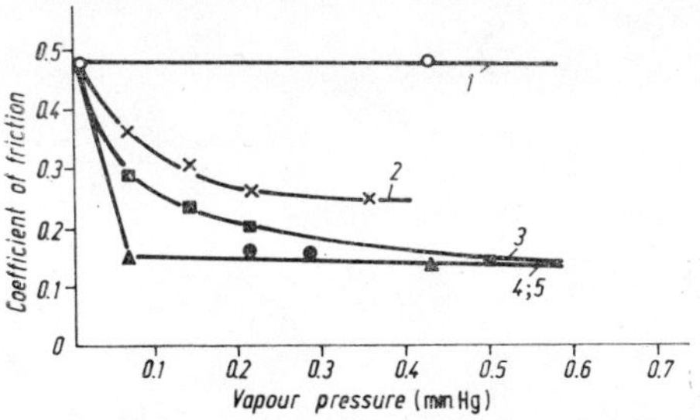

Figure 4.24 Effect of atmospheric composition and pressure on the coefficient of friction of graphite [150]:

1, azote; *2*, hydrogen; *3*, water vapours; *4*, oxygen; *5*, heptane.

vapour required to reduce the coefficient of friction to its normal value is very small, adsorption is considered to take place only on the crystal edges. The adsorbed layer reduces the friction between crystallites allowing their more favourable orientation [40, 54].

The use of graphite in solid film form is restricted by its relatively low thermal stability and the variation of the coefficient of friction with the gaseous atmosphere. However, the use of graphite in the form of either a plain or colloidal suspension has solved many lubrication problems under conditions of severe pressure and temperature.

4.3.2 **Molybdenum disulphide**

Molybdenum disulphide has a hexagonal-type crystal structure in which each atom of molybdenum is surrounded at equal distances by 6 atoms of sulphur placed at the corners of a triangular right prism of 3.17 Å height. Each layer consists of 2 planes of sulphur atoms and an intermediate plane of the molybdenum atoms. The distance between the planes of molybdenum and sulphur atoms is of 1.58 Å (Figure 4.25). The bonds between the atoms of molybdenum and

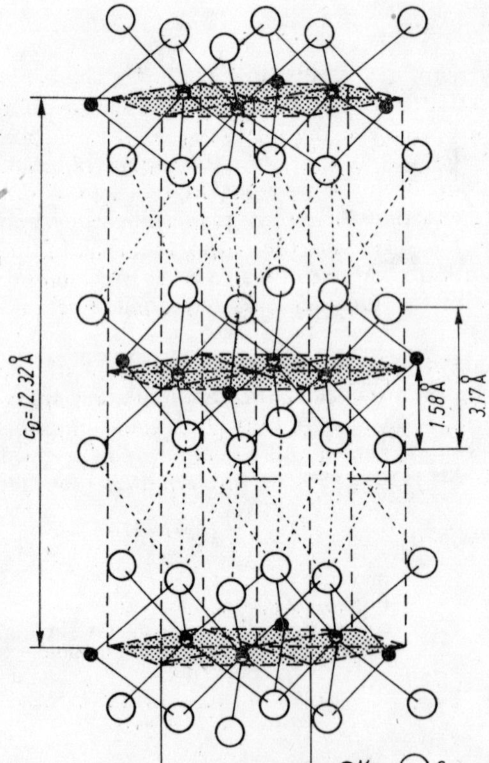

Figure 4.25 Structure of molybdenum disulphide.

sulphur are covalent while between the atoms of sulphur the bonds are van der Waals type. The low binding energy between the planes of the sulphur atoms allows molecular sliding along these planes. The shearing force further decreases due to lattice defects present in real crystals.

In addition to the hexagonal structure, a rhombohedral crystal structure is possible in high purity synthetic molybdenum disulphide [18].

The properties of molybdenum disulphide are affected by several factors of importance to its performance as a solid lubricant and these have been reviewed (Winer, Barz) [181, 24]. In the present discussion only the essential aspects of the properties of the MoS_2 layer will be covered.

4.3.3 **Effect of humidity**

As distinguished from graphite, molybdenum disulphide owes its beneficial frictional characteristics directly to its crystal structure and not to the presence of an adsorbed layer on the microcrystal surface. The coefficient of friction of MoS_2 decreases as the pressure decreases and the purer the MoS_2 used, the lower is the coefficient of friction.

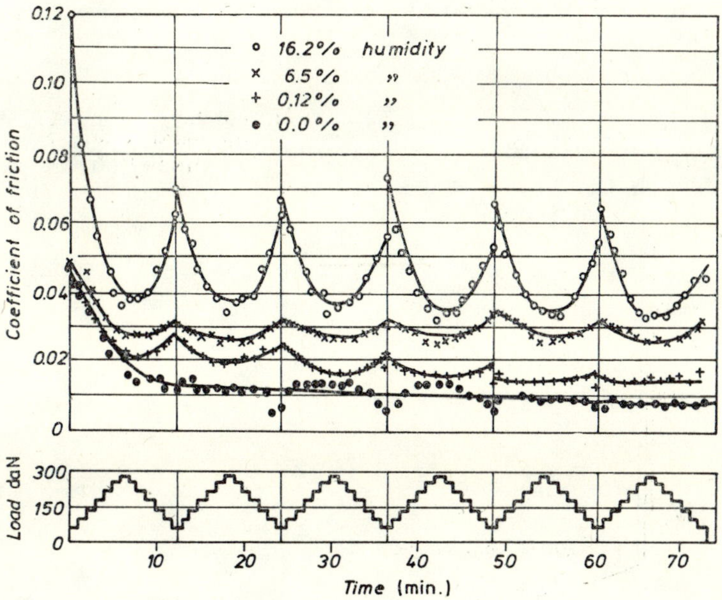

Figure 4.26 Effect of humidity on the coefficient of friction for MoS_2 [64].

Under normal atmospheric conditions the concentration of water vapour has a significant effect on the coefficient of friction of MoS_2 [138, 64]. Figure 4.26 shows the variation of the coefficient of friction of MoS_2 with load for various relative humidities. Besides the direct effect of humidity, the minimum values on the friction coefficient curves are considered to be the result of dehydration of the film due to a temperature rise on the frictional surface. This hypothesis is supported by the fact that in a dry atmosphere (0% humidity) the effect of load is considerably reduced. An increase in speed produces a similar decrease in the coefficient of friction but only in a wet atmosphere (Figure 4.27).

Humidity and atmosphere oxygen content greatly affect film endurance. Considering film endurance as the duration of a constant low coefficient of friction de Gee and Solomon [158] determined the effect of humidity and oxygen content

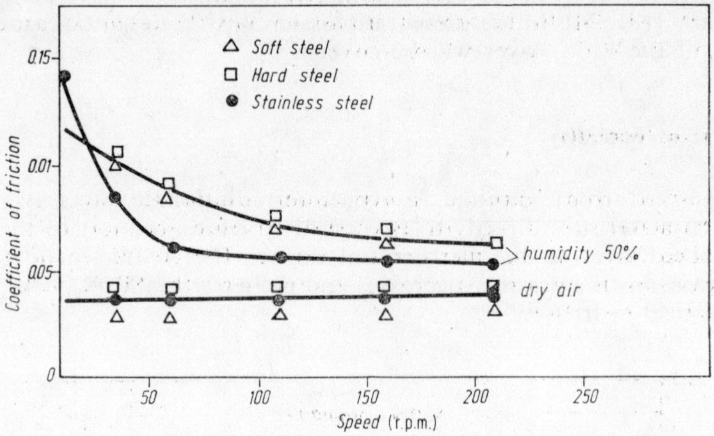

Figure 4.27 Coefficient of friction of MoS_2 film versus speed [181].

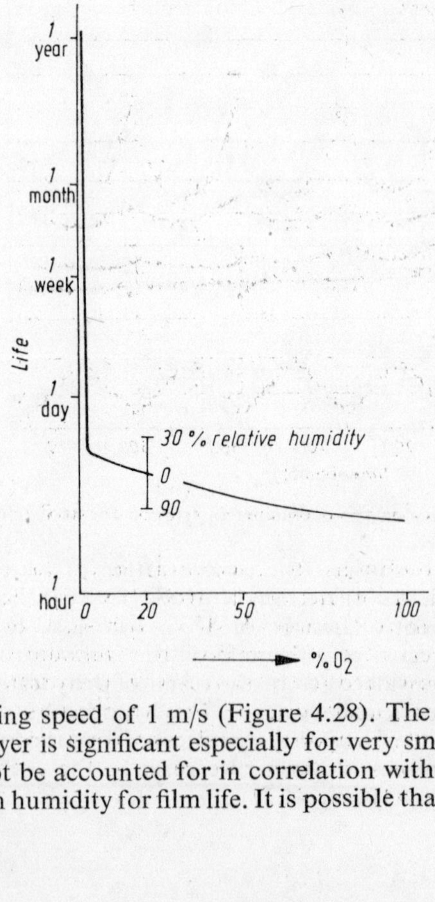

Figure 4.28 Wear life of MoS_2 film versus oxygen content and relative humidity [158].

at a sliding speed of 1 m/s (Figure 4.28). The effect of oxygen, due to oxidation of the layer is significant especially for very small concentrations. An effect which could not be accounted for in correlation with other results is the existence of an optimum humidity for film life. It is possible that when a film is formed by frictional contact with MoS_2 powder the presence of water vapour allows a thicker film to be obtained by means of better orientation of the microcrystals [97].

The role of water vapour and oxygen may be explained by the existence under normal atmospheric conditions of a monomolecular layer of molybdenum trioxide on the MoS_2 crystals. This layer which, as distinguished from MoS_2 is hydrophilic reduces the energy of adhesion between the crystals and the support surface and hence the film life. The monomolecular layer of MoO_3 protects the crystal surface against further oxidation up to a temperature of 370°C [31].

If MoS_2 is heated in an atmosphere which is low in oxygen or in a carbon dioxide atmosphere, molybdenum dioxide is formed which seems to have abrasive properties [77]. Oxidation of MoS_2 in the presence of carbon dioxide with the formation of MoO_2 is important in practical applications such as a solid lubricant in nuclear reactors cooled with carbon dioxide.

The contact surface of the friction pair has a considerable effect on the coefficient of friction of MoS_2 films. With concentrated load the coefficient of friction is lower than with surface contact. As distinguished from MoS_2, other similar type compounds, such as WSe_2, have a low coefficient of friction even at low pressures.

4.3.4 Effect of temperature

Three temperature ranges may be considered in the examination of the effect of temperature on the friction of MoS_2 films, namely: a range of low temperatures up to 0°C, a range of normal temperatures from 0°C to 100°C, and a range of high temperatures over 100°C. In the low temperature range which is important in rocket engines, available data do not show any significant difference in behaviour compared with that at normal temperatures. What is more important under low temperature conditions is that MoS_2 shows no tendency of self-ignition when it comes into contact with liquid oxygen.

In the normal temperature range the coefficient of friction decreases as the temperature increases due to reduction of the water content of the film. If measurements are carried out in a dry atmosphere, the coefficient of friction remains constant over a large temperature range.

Over the high temperature range the wear behaviour of MoS_2 film is significantly affected by the chemical changes which take place. In the presence of oxygen or water vapour the maximum operating temperature is considered to be 450°C. In an atmosphere of hydrogen the maximum operating temperature increases to 700°C, and in vacuum up to 1100°C. Sublimation of the layer occurs at 1300°C while at 1600°C in vacuum, thermal dissociation with the formation of metallic molybdenum takes place.

4.3.5 Effect of vacuum and ionizing radiations

In contrast with graphite the coefficient of friction of MoS_2 films decreases with the decrease in pressure. For a vacuum of the order of 10^{-6} torr a decrease of the coefficient of friction in operation and a return to its initial value in static conditions has been observed. It is considered that a removal of the molecules retained on the film surface occurs by friction [98].

As MoS_2 is used in some mechanisms incorporated in nuclear reactors its resistance to high radiation doses is particularly important. In most cases, this solid lubricant is used together with a binder in the form of mixed films. The life of such films has been established by measuring the coefficient of friction in

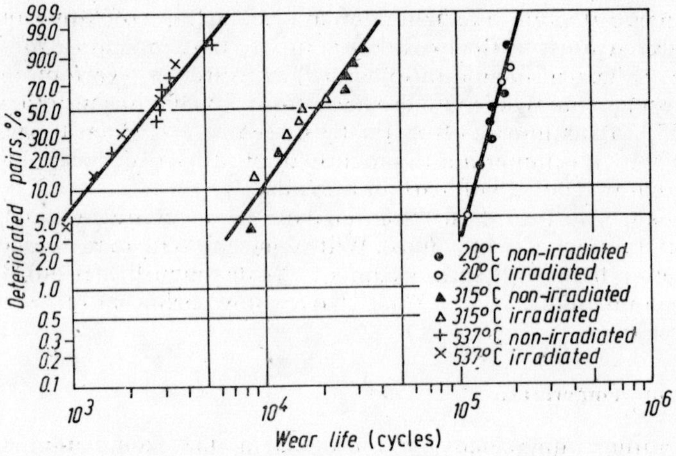

Figure 4.29 Effect of radiations on wear life of MoS_2 + PbS + + B_2O_3 film at various temperatures [52].

terms of the number of rotations for various compositions, temperatures and radiation doses.

Figure 4.29 shows the life of a mixed film of MoS_2 + PbS + B_2O_3 on a steel support expressed in percent of the deterioration of friction pairs. The pairs were subjected to mixed irradiation of approximately $2.2 \cdot 10^{-11}$ erg/gr γ radiations and an additional irradiation by neutrons. The radiation dose used does not appear to affect the wear behaviour of the film.

4.3.6 **Effect of support surface**

When considering the effect of the support surface account should be taken of the following factors: the surface material, hardness and roughness. Although available research data allow no general conclusions to be drawn, they have indicated the significant effect of the nature of the surface on film life. Table 4.2 shows the effect of the material of the friction pair on film life. The measurements were carried out with a pin on plate type apparatus under a total load of 20 daN and at a sliding speed between 0.6 and 1 m/s.

Under the test conditions used in Table 4.2 a friction pair of steel on steel has an endurance of the order of several hours. If one part of the friction pair is made of aluminium or titanium, the endurance is considerably reduced whereas for a friction pair of titanium on titanium the endurance is of only several seconds duration.

Surface hardness is another parameter which affects film endurance. Figure 4.30 shows the effect of surface hardness on film endurance, expressed as the number of rotations. Results have been obtained for various materials with various hardness, so that the effect of the surface nature could be consequently taken into account.

Table 4.2

Effect of support material on molybdenum disulphide film life [158]

No.	Pair of materials	Life
1	Steel — glass	days
2	Titanium — glass	days
3	Steel — steel	10—20 hours (*)
4	Steel — aluminium	minute up to hours (**)
5	Steel — titanium	minutes (***)
6	Titanium — titanium	seconds (***)

(*) Function of humidity.
(**) Function of load.
(***) Even for light loads.

By expressing film endurance in terms of surface roughness an optimum value for this parameter was found. Thus, for $R_a = 0.3 \div 0.4$ μm the film endurance was approximately 13 hours, while for a roughness $R_a = 0.04$ μm the endurance was only several minutes [158]. A higher roughness also leads to a lower endurance. The existence of an optimum roughness may be accounted for by the fact that at $R_a = 0.3 \div 0.4$ μm the thickness of the initially deposited film is a maximum.

The direction of surface processing also has an effect. The coefficient of friction is minimum when the sliding direction is parallel to the processing direction and maximum if the two directions are normal to each other.

With the same surface support material film endurance may be increased by certain thermochemical treatments. Such treatments, which form a new surface

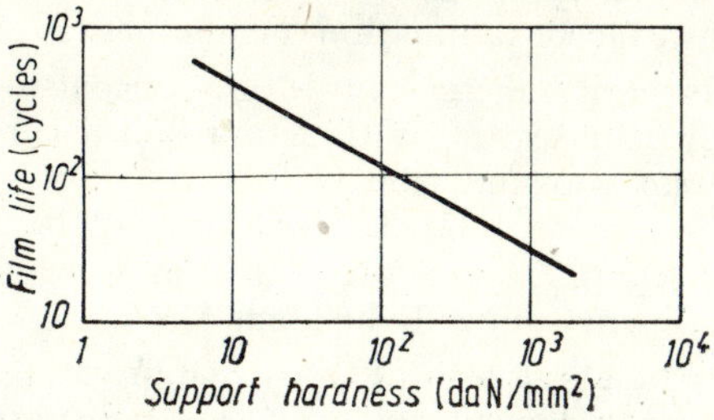

Figure 4.30 Wear life of MoS_2 film versus support surface hardness [106].

increase the adhesion of the MoS_2 film and hence its endurance. Figure 4.31 shows the effect of surface treatment on the endurance of the MoS_2 film achieved by means of a paste. A maximum endurance of about 10 times greater than that of untreated surfaces has been obtained by the use of sulphonitriding.

The effect of thermochemical treatments on the coefficient of friction may be either positive or negative. This does not affect the applications of such treatments as it is endurance that is most important for the application of solid films in practice.

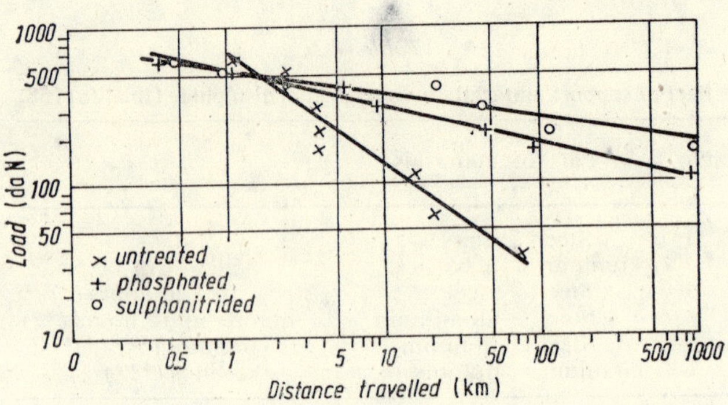

Figure 4.31 Effect of surface treatment on wear life of MoS_2 film [123].

4.3.7 Molybdenum disulphide as a lubricant additive

Molybdenum disulphide may also be used as an additive in mineral oils or in lubricating greases [76, 55]. Although the effectiveness in these instances is also due to the MoS_2 layer, the presence of the lubricant introduces new parameters: the concentration of molybdenum disulphide, the particle size and the interaction between MoS_2 and other chemical additives present in the lubricant.

As under such conditions the surfaces of the friction pair are in constant contact with the MoS_2 suspension, the efficiency of the product is measured by means of surface wear, seizure load and the coefficient of friction. As the layer of solid lubricant formed on the frictional surfaces with an MoS_2 suspension is much thinner than if the layer were pure, its presence can only be established by special surface examination techniques.

Generally, increase in MoS_2 concentration in the lubricant has a beneficial effect regarding wear. With a four-ball tester it was determined that the diameter of the wear scar under heavy load tests diminishes as the concentration of MoS_2 increases (Figure 4.32). Under light load the concentration of MoS_2 has no effect (Göttner) [76]. It was found that in some cases an optimum concentration of MoS_2 occurred under light loads [161].

Particle size also affects the lubricating properties of the suspension. Experimental work carried out on a four-ball tester has shown that if colloidal suspensions are used, the optimum mean diameter of the particle is 1.5 μm.

Other authors found a more complex interdependence between particle size and the anti-wear characteristics of the suspension. Thus, under light loads particle size has no effect, while under heavy loads larger grain sizes result in increased wear [76]. The interdependence between the above factors is further complicated by classifying the suspensions by the mean particle size, as various distributions of particle diameter may exist for the same mean size.

MoS_2 is usually used in fluid lubricants together with other additive which may aid or prevent its action. As some of the additives are surface-active substances

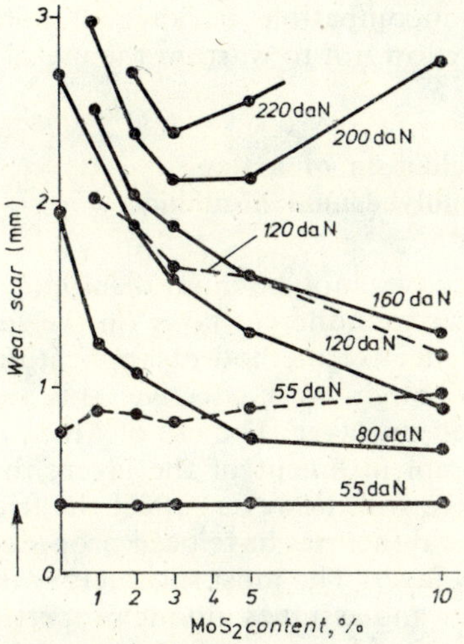

Figure 4.32 Wear determined on a four-ball machine function of MoS_2 concentration in the lubricant (Load as parameter) [76].
SAE 90 oil; -- operating time: 1'; – – operating time: 1h.

they form an adsorbed layer on the metal surface, a layer which may reduce adhesion or even prevent the formation of a MoS_2 film. This effect has been shown by using a suspension of MoS_2 in pure hydrocarbons and pure hydrocarbons containing polar compound as additives. Maximum effectiveness was obtained with MoS_2 and a pure lubricant. The adverse effect of polar substances may be counteracted by the use of high concentrations of solid lubricant [63]. The importance of the above effect is of significance when molybdenum disulphide powder is added

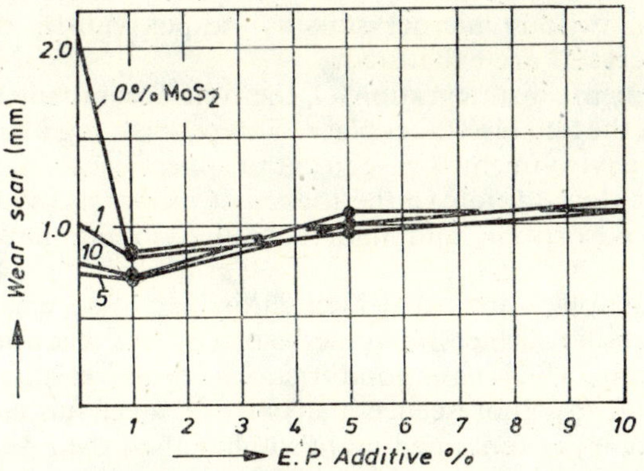

Figure 4.33 Effect of extreme pressure additives on MoS_2 efficiency in gear lubricants [23].

to an additive fortified lubricant, such as a lubricant containing detergent or extreme pressure additives.

The effect of MoS_2 additives on gear lubricants was studied with a four-ball and a gear test machine (Figure 4.33). Under the test conditions a beneficial effect was found only up to a concentration of 1% E. P. additives. With the gear test machine use of molybdenum disulphide as an additive may sometimes even increase wear [23].

It may be concluded that, in contrast to preformed molybdenum disulphide layers, the effectiveness of suspensions should be carefully considered before use as, if an incompatible package of additives is used, the effectiveness may be sufficiently low not to warrant the use of MoS_2.

4.3.8 Mechanism of action of molybdenum disulphide

To be effective, molybdenum disulphide should be able to reach the loaded area and to form an adherent layer on the surfaces of the friction pair. Consequently, the most effective method of application is to form a pure film on the frictional surface. Though most effective, this technique generally does not allow replenishment of the layer. The use of MoS_2 as a lubricant additive allows the possible constant replenishment of the layer; the presence of the lubricant and any chemical additive will, however, affect the frictional characteristics of the layer.

Several theories have been proposed to explain the good frictional properties of a MoS_2 layer. The most satisfactory theory is based on the crystalline structure of the layer and assumes unique properties of the layer due to its lamellar structure which causes anisotropy of the crystallites regarding their mechanical properties. Under the simultaneous action of the normal and tangential forces, displacement occurs between the various planes of sulphur atoms, the last of them being attached to the moving surface. This theory is also supported by the fact that the worked MoS_2 layer is highly oriented with the planes of sulphur atoms parallel to the surface.

The image of the package of lamellae sliding relative to one another is valid only for a single microcrystal. The MoS_2 film on the frictional surface is however made up of many superposed microcrystals. The macroscopic tangential displacement occurs successively on various microcrystals as the possibilities of tangential displacement in a microcrystal are exhausted.

Simultaneously with displacement, disintegration of the film accompanied by separation of the microcrystals in the film also takes place. This process, together with the chemical reaction with the environmental gaseous atmosphere eventually leads to the destruction of the layer. In addition to the theory of intracrystalline displacement, theories of gaseous adsorption and intercrystalline sliding have also been proposed [80, 54].

Besides the mechanical interaction, a chemical interaction may also exist between the MoS_2 layer and the metallic surface. In some instances, the reaction products are themselves solid lubricants which have good frictional characteristics. This effect has been established by X-ray fluorescence analysis of a steel surface lubricated with MoS_2. As the surface layer contained more sulphur than the layer corresponding stoichiometrically to molybdenum disulphide, it was concluded that this was due to the presence of FeS. The reaction of MoS_2 with iron was established both statically in a mixture of iron powder heated at temperatures over 720°C and dynamically on an Almen-Wieland machine in which the required temperature was achieved by the energy dissipated in the friction pair [103].

4.3.9 Techniques of application of molybdenum disulphide

Due to its outstanding properties, molybdenum disulphide is a commonly used solid lubricant. The variety of applications for which it has been used has resulted in numerous techniques for its use to utilize fully its beneficial properties.

Dry Films. Such films are formed by friction of metallic surfaces with molybdenum disulphide powder. The thickness of the resulting layer, which determines directly the life of the film depends, among other factors, upon the pressure used and the running time. An essential condition for good adhesion is thorough cleaning of the surface to remove any traces of oil, rust, dust, etc. Dry films, which have a limited life, are used mainly for surface treatment prior to assembly and after repair, to lubricate some parts subjected to discontinuous motion, under hostile conditions such as high temperature.

To coat surfaces not readily accessible suspensions of MoS_2 in volatile liquids which evaporate after spraying have been used. The film obtained is however less adherent and not oriented. Concentrates of MoS_2 in paste form are sometimes used instead of the powder and the paste can be readily applied with a brush.

An effective method of continuous layer renew is to introduce tablets of MoS_2 into the frictional contact. Their passage over the loaded area regenerates the solid lubricant film.

Varnishes. The incorporation of MoS_2 in synthetic resins or other binders represents an advance in the development of dry films. By spraying, immersion, painting, etc., of such suspension of MoS_2 it is possible to obtain adherent films of various thicknesses on any surface. The binder should be carefully selected so as not to affect the qualities of the molybdenum disulphide. Furthermore, the binder should be elastic, anticorrosive, temperature resistant and chemically inert. The effect of the binder and other factors on the frictional characteristics of the resulting varnish is shown in Table 4.3. As with dry films, adequate surface treatment prior to application of the varnish improves film life. The most commonly used film thicknesses are between 0.01 and 0.02 mm.

Composite materials. Another widely used technique for the application of MoS_2 is to incorporate it in the so-called composite materials. By selecting adequate compositions and by compaction of the resulting mixture, the composite materials obtained offer new materials with predicted properties. Self-lubricating materials are an important class of composite materials. The lubricant component is a solid lubricant usually molybdenum disulphide or a similar substance such as graphite. The self-lubricating composite material can be used for the manufacture of certain parts or to coat metallic surfaces which are required to operate under unlubricated conditions.

Suspensions of MoS_2. Greases. Molybdenum disulphide may be used in relatively low concentrations ($<1\%$) as an additive in gear and engine lubricants, and in somewhat higher concentrations in cutting oils. As shown in § 4.3.7 efficiency is affected in this instance by the presence of the lubricant as well as by the presence of chemical additives.

Of the chemical additives emulsion stabilizers are important as they keep the MoS_2 particles in suspension. Such stabilizers, which are also detergents, should be carefully selected to avoid adversely affecting the process of film formation. Suspensions of MoS_2 have been successfully used as additives in gear and engine lubricants and during the running-in of large gears.

Table 4.3

Wear characteristics and the coefficient of friction for the resin-MoS_2 systems

Test	Resin content: MoS_2 content	Type of resin: silicone	phenolic	epoxi	epoxi 60%, polyamide 40%
Wear scar diameter (mm) (four balls)	1: 1	1.1	1.5	1.3	1.4
	1: 3	1.4	1.2	1.1	0.95
	1: 9	0.53	0.62	0.43	0.39
Mean Herz Load (daN)	1: 1	50	37	73	51
	1: 3	82	78	50	55
	1: 9	74	122	60	101
Coefficient of friction	1: 1	0.25	0.17	0.12	0.12
	1: 3	0.12	0.11	0.12	0.11
	1: 9	0.13	0.05	0.10	0.07

Molybdenum disulphide, however, is now mainly used as an additive in greases in concentrations of 0.5 ÷ 18%. The most popular MoS_2 containing greases are lithium soap base greases. The role of the solid lubricant is to protect the frictional surfaces if the lubricating capacity of the base grease is exceeded due to thermal or mechanical stresses. The addition of molybdenum disulphide to greases has proved to be effective in the following cases [200]:

(a) the lubrication of journal bearings at temperatures close to or higher than the pour point. The optimum concentration of MoS_2 is 6–8%.

(b) the lubrication of heavily loaded ball bearings. In this case, the optimum concentration of MoS_2 is low ($\leqslant$1.5%), while the average grain size of the MoS_2 should be $<5\ \mu m$.

(c) the lubrication of surfaces at temperatures above 250°C. In this case, the grease acts only as a carrier for the MoS_2.

4.3.10 New materials similar to molybdenum disulphide

Studies of the properties of some sulphides, selenides and tellurides have led to the recent discovery of new lamellar solid lubricants, most of which crystallize in the hexagonal system. Among the metals used are: wolfram, vanadium, niobium, tantalum, and hafnium. Comparison of the characteristics of some of these compounds with graphite and molybdenum disulphide under conditions of high temperatures and high vacuum are shown in Figure 4.34. Although molybdenum disulphide is superior to graphite under such conditions, it has a tendency to release sulphur due to the destruction of the crystalline lattice. This behaviour reduces the film frictional properties and contaminates the apparatus in which it is used.

The oxidizing stability in the air of the new compounds is also superior to that of graphite and molybdenum disulphide, as shown in Figure 4.35. It has been found that similar compounds of thorium and vanadium raise the maximum operating temperature to over 1900°C.

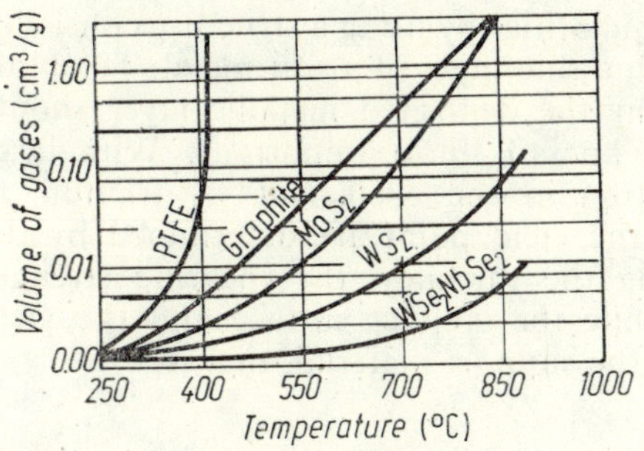

Figure 4.34 Volume of gas released by several solid lubricants at a pressure of 10^{-2} torr versus temperature [114].

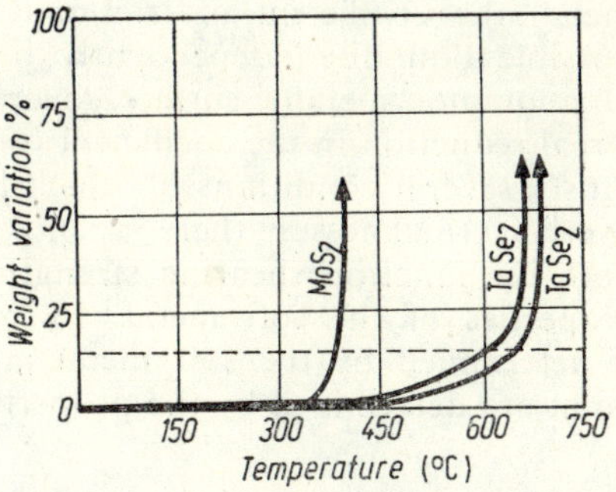

Figure 4.35 Resistance to oxidation of compounds with tantalum compared with MoS_2 [114].

Except for the tellurides, the coefficient of friction of the new compounds are lower than those of graphite or molybdenum disulphide. With niobium, tantalum and wolfram, this characteristic may be accounted for by the larger atomic radii of these metals as compared with molybdenum. An increase in the atomic radius leads to a weakening of the van der Waals-type bonds between the cleavage planes, which affects the shearing force in a direction parallel to these planes.

Certain selenides and tellurides, due to their good frictional characteristics and high electrical conductibility allowed the satisfactory functioning of moving electrical contacts even under high vacuum conditions.

Figure 4.36 Dimensional stability to irradiation of several composite materials. Irradiation dose: gamma 10^9Rö; neutrons 10^{17} n/cm^2.

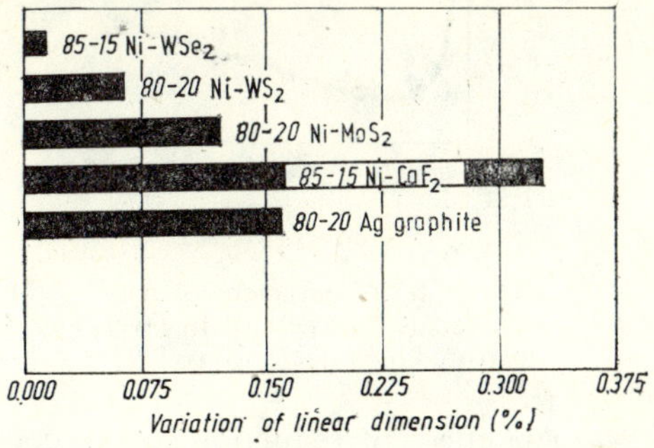

These materials are usually incorporated in composite materials to ensure a satisfactory mechanical strength. The new types of composite materials have allowed extension of the operating temperature and vacuum ranges and of the radiation dose beyond the limit of conventional materials (Figure 4.36).

4.4 Soft metallic layers

According to relation (2.6), in metallic contact, the coefficient of friction is directly proportional to the shearing strength of the junction and inversely proportional to the yield point of the material of the friction pair. As both quantities vary simultaneously in the same direction depending on the nature of the metal, the coefficient of friction is constant within known narrow limits. It is possible to modify independently the two quantities by using a structure consisting of a common metallic surface covered with a thin layer of a soft metal. To obtain a real reduction in the coefficient of friction, the deposited metallic layer should satisfy several conditions of which the thickness is most important. With large overlay thicknesses there is no reduction in the coefficient of friction as both the junction shearing strength and the yield point are determined by the properties of the soft metal. By reducing the thickness the shearing strength is determined by the soft metal layer while the contact surface and the yield point are determined by the properties of the support material, that is

$$\mu = \frac{(\tau)_{\text{layer}}}{(\sigma_c')_{\text{support}}}. \tag{4.13}$$

By further reducing the thickness of the deposited layer penetration of the layer occurs and the coefficient of friction again increases (Figure 4.37). Also, a thin overlay is detrimental to the life of the layer.

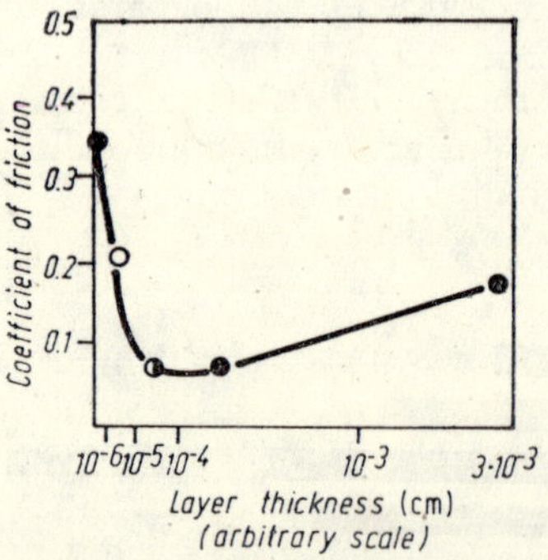

Figure 4.37 Coefficient of friction versus thickness of In layer deposited on a steel surface [39].

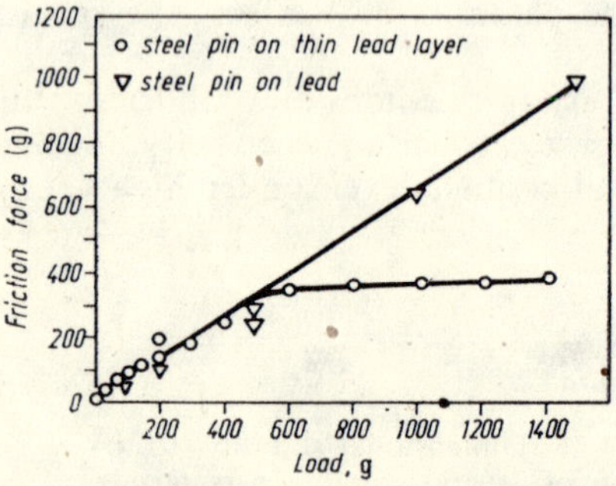

Figure 4.38 Friction force versus load for a thin layer of lead and massive lead [46].

The effect of reduction in the coefficient of friction for a hard support—soft metallic layer structure may also be obtained for a given layer thickness by increasing the load (Figure 4.38). Under light loads there is no difference in the coefficient of friction of a solid lead surface and that of a steel surface covered with lead. Under a certain load, however, a metallic layer on a steel supporting surface reduces the coefficient of friction.

For thin metallic layers to be effective several other conditions should be fulfilled. They include good adhesion of the film to the support surface, compatibility between the film and the counterpart material and a low layer wear rate. The correct properties should always be selected depending on the specific operating conditions of the friction pair.

The adhesion to the support surface and the compatibility (the tendency to alloying) with the counterpart material are two contradictory requirements if the two parts of the friction pair are of the same material. If a metal which is soluble in the support surface material is chosen to provide good adhesion, the layer will consequently tend to adhere to the counterpart. To avoid this one or several intermediate layers are deposited. If the friction pairs are steel (S) and a metal (A) insoluble in steel is chosen to form the layer, then an intermediate layer B is used to ensure good adhesion of the layer A to the steel, i.e. the arrangement $S/B/A/-S$ is obtained. Metal B is soluble in both steel and metal A. It is possible, however, that metal B may diffuse through metal A during operation, thus coming into contact with the steel surface and causing seizure so, a system of three layers, namely $S/B/C/A-S$ is used, where C is soluble in the metals B and A but insoluble in steel. A continuous chain of solid solutions is thus obtained.

The metals selected to form layers should also fulfil certain conditions in order to ensure a satisfactory mechanical resistance of the whole system, namely:

— the coefficient of expansion of the metals forming the layers should increase outwards from the base,

— the gradient of the mechanical properties should be positive inwards from the surface.

It was found that the system Steel/Ni/Au/Ag is most satisfactory for steel. The layers are deposited successively by electrolysis, the outer silver layer being the thickest (5 μm) [192]. The coefficient of friction for such a system, determined on a Faville machine, under a load of 150 daN in the presence of demineralized water was $\mu \simeq 0.11$. This value remained constant for about 3 min.

The range of application of multiple layers involves low sliding speed pairs and discontinuous running pairs which are inadequately lubricated or which operate in a hostile environment.

High vacuum is another field in which thin metallic layers are used. Conventional alloys employed in mechanical engineering cannot be used in vacuum due to the adhesion and wear which they suffer. Such surfaces can be lubricated with soft metallic layers. Figure 4.39 shows the effect of the nature of the layer and the method of deposition on the coefficient of friction and wear rate in air and vacuum. It should be noted that the coefficient of friction in air for the steel surface covered with a lead layer is higher than that for the steel surfaces. This behaviour is probably due to the use of too thick a lead layer.

A development of thin metallic films is the simultaneous deposition of a ductile and a heat-resistant metal in a mixed layer form. The soft component acts as a binder, while the heat-resistant, readily oxidizable metal acts as a lubricant at high temperatures. Such mixed metallic layers (Au — Re or Au — Mo) of 0.01 to 0.02 mm thicknesses deposited on a steel support have operated satis-

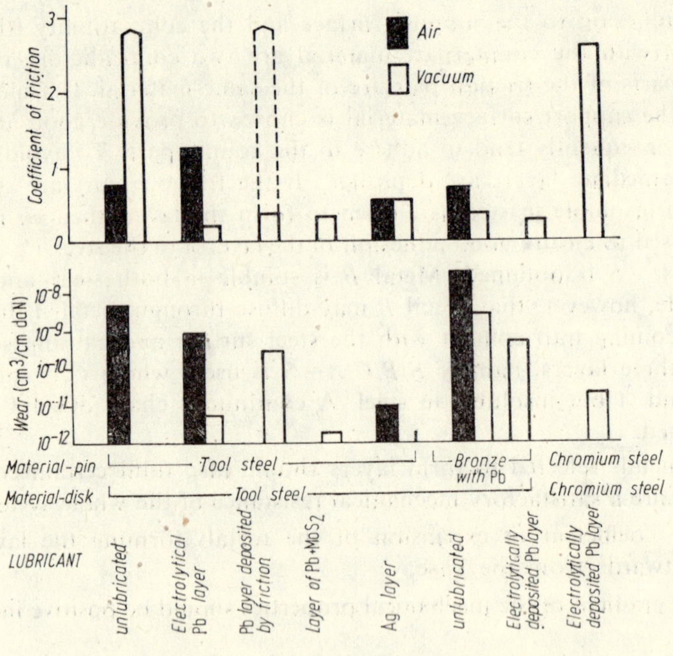

Figure 4.39 Coefficient of friction and intensity of wear in air and vacuum for several materials and metallic layers [81].

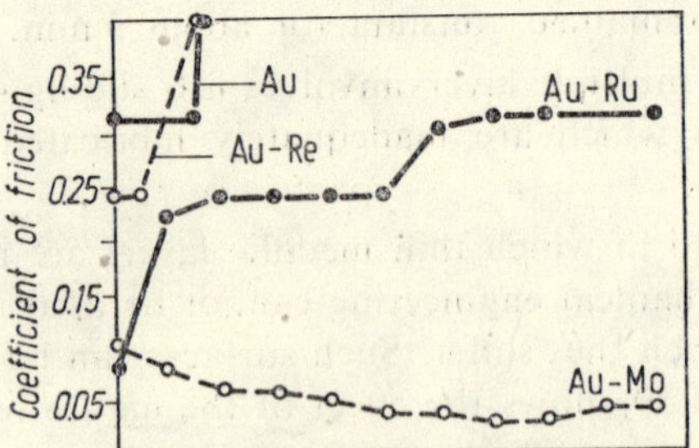

Figure 4.40 Film life for several mixed metallic layers at 500°C [170].

factorily even above 500°C. Figure 4.40 shows the operating time and the coefficient of friction for the combinations Au—Re, Au — Ru, Au — Mo. The Au — Mo layer still has a coefficient of friction of $\mu < 0.05$ after 3 000 passages.

5 Friction of metals and alloys in air and controlled atmosphere

The friction and wear of metallic pairs in air, controlled atmosphere and vacuum are considerably affected by the processes of metallic surface oxidation, adsorption and chemisorption of gases or vapours. In all cases except ultrahigh vacuum the metallic surfaces of the friction pair are covered with oxides and adsorbed gases which fundamentally change the frictional properties of the surfaces. The presence of the oxide layer on the surfaces of the pair is the result of a dynamic process of formation on the one hand, and wear on the other hand. As distinguished from oxidation under static conditions, oxidation of the frictional surface is stimulated by both the existence of a high temperature and the activation of surfaces as a result of the frictional process.

5.1 Oxidation of metals and alloys

Over a large temperature range all metals except gold react with oxygen to form oxides. In the equilibrium state, the resulting compounds are a function of the oxygen pressure and temperature. The existence of a certain compound is however conditioned also by its rate of formation. At low temperatures certain oxides have such a low formation rate that the process practically does not even occur. The formation of oxides by the reaction of oxygen with metals is a typical surface reaction. The reaction product, namely the oxide, separates the reactants thereby reducing the rate of reaction. Depending upon the constitution of the oxide layer, which may be continuous, porous or cracked, continuation of the reaction is ensured by oxygen diffusion into the layer or by the penetration of molecular oxygen through the pores. An inverse diffusion of metal atoms also takes place. In the case of diffusion through a homogeneous solid layer the transport of the material occurs by means of lattice or interstitial defects. The quantity of solute m diffused through a surface A in unit time is proportional to the concentration gradient $\mathrm{d}C/\mathrm{d}x$ considered to be normal to A

$$m = -AD\frac{\mathrm{d}C}{\mathrm{d}x}. \tag{5.1}$$

Variation of the concentration C with time at the point where the concentration gradient is dC/dx, if D is constant, may be written as

$$\frac{\partial C}{\partial t} = D \cdot \frac{\partial^2 C}{dx^2} \tag{5.2}$$

The dependence of the diffusion coefficient on temperature is given by an Arrhenius-type relation

$$D = D_0 e^{-\frac{Q}{RT}}, \tag{5.3}$$

where D_0 and Q are constants.

The values of D_0 and Q for several oxides are listed in Table 5.1.

Table 5.1

Diffusion constant D_0 and activation heat Q for several oxides (after [105])

Compound	Diffusing ion	Temperature range °C	D_0 ($cm^2 \cdot s^{-1}$)	Q (kcal/g)
FeO	Fe	690–1010	0.014	30.2
Fe_3O_4	Fe	750–1000	5.2	55
Fe_2O_3	Fe O	930–1270 1150–1250	$4 \cdot 10^4$ 10^{11}	112 146
Cu_2O	Cu O	800–1050	0.12 0.0065	36.1
Cr_2O_3	Cr	1000–1350	4000	100
Al_2O_3	O O	1200–1600 1500–1780	$6.3 \cdot 10^{-8}$ 1900	57.6 152

Due to the specific conditions existing on the surface certain compounds which are unstable in bulk at the same temperature may be formed. For instance, by the oxidation of iron a layer of FeO is formed and preserved under the layer of Fe_3O_4 even at temperatures below 570°C, although FeO is unstable below this limit.

If the metal forms several oxides then these compounds are arranged so that the compound which is richest in oxygen lies at the solid-gas interface while the compound which is poorest in oxygen lies at the metal-oxide interface (Figure 5.1). If an alloy is subjected to oxidation, the reaction products occur either in the form of separate layers or as a heterogeneous mixture. In the latter case, a new mixed compound may form. If one of the oxides is in excess it may impose its structure on other compounds present in smaller quantities. The structure of the main compound may be affected by the structure of the base metal.

Several properties of the oxides and hydroxides of some metals used in mechanical engineering are listed in Table 5.2. The latent heat of transformation with the passage of the oxide to a lower oxidation state and formation of a mole-

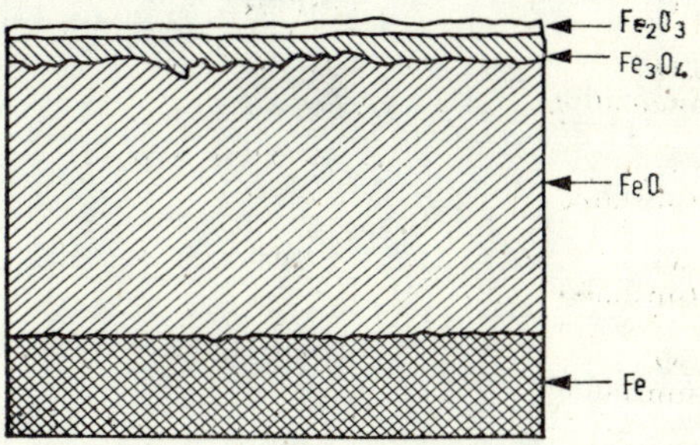

Figure 5.1 Formation of superposed oxide layers on an iron surface at 625°C after 24 h [105].

cule of oxygen has been denoted by ΔH_{dec}. The last column gives the ratio φ between the molecular volume of the oxide and the atomic volume of the metal

$$\varphi = \frac{\text{molecular volume of oxide}}{\text{atomic volume of metal}} . \tag{5.4}$$

The magnitude of this ratio may be taken as a measure of the stresses occurring in thin oxide layers or as an indication of the tendency for crack formation in thick oxide layers. The greater the value of φ for an oxide, the smaller the thickness cracks occur.

The first stage in the reaction between a metal and oxygen is the chemisorption of oxygen. As the rate of chemisorption is very high, at normal temperatures the surface is almost instantaneously covered with a monomolecular layer of oxygen. The oxide layer proper is formed by the complete reaction between this oxygen layer and atoms in the metal surface. Then, up to a thickness of several hundred Angström the increase in the thickness of the oxide layer takes place by the diffusion of the reactants in both directions. With greater thicknesses of oxide pores or cracks usually occur, facilitating access of oxygen to the metallic surface.

An important characteristic of the oxidation process is its evolution as a function of time. If the mass variation per unit area of the sample Δm is taken as a measure of oxidation, then the kinetics of the oxidation process is given by the relation

$$\Delta m = f(t). \tag{5.5}$$

Depending on the thickness of the layer and temperature, relation (5.5) may be a linear, parabolic, cubic or logarithmic function. For thin oxide layers

Table 5.2

Several properties of oxides (after [105])

Compound		Structure	Melting point (°C)	H_{dec} (kcal)	Molar volume (cm^3)	φ
FeO		NaCl type	1 424	126.5	11.9/12.5	—
Fe_3O_4		spinel	1 597	144.5	44.7	2.10
Fe_2O_3 (hematite)	α	Cr_2O_3 type	675 *	110.2	30.4	2.14
Fe_2O_3 (unstable)	β	β-Al_2O_3 type	—	—	—	—
Fe_2O_3 (unstable)	γ	cubic	—	—	—	—
Fe_2O_3 (unstable)	δ	hexagonal	—	—	—	—
FeOOH	α	orthorhombic	230 *	18	21.3	3.0
FeOOH (unstable)	β	orthorhombic	—	—	27.5	—
FeOOH	γ	orthorhombic	400 *	22.5	22.4	—
$Fe(OH)_2$		$Cd(OH)_2$ type	—	13.0	26.4	3.7
$Fe(OH)_2$ (unstable)		amorphous	—	—	—	—
Cu_2O		cubic	1 230	80.0	23.3	1.64
CuO		monoclinic	1 100 *	68.4	12.2	1.72
$Cu(OH)_2$			35 *	10.5	29.0	4.0
Al_2O_3 (corundum)	α	Cr_2O_3 type	2 030	266.8	25.6	1.28
Al_2O_3 (unstable)	β	hexagonal	—	—	30.7	1.54
Al_2O_3 (unstable)	γ	spinel	—	253.1	29.8	1.49
Al_2O_3	δ	tetragonal	—	—	—	—
Al_2O_3 (results from anodic oxidation)		cubic	—	—	—	—
AlOOH	α	α-FeOOH type	—	—	17.8	1.78
AlOOH	γ	α-FeOOH type	300 *	5.2	19.5	1.95
$Al(OH)_3$	α	amorphous	—	12.3	30.9	3.09
$Al(OH)_3$	γ	monoclinic	150 *	12.9	32.0	3.2
Cr_2O_3		rhombohedral	2 400	180.0	29.0	2.07

* Transformation or decomposition heat

a logarithmic or cubic function is valid in most cases, whereas for thick oxide layers a parabolic or linear function is valid. The oxidation-time curves at room temperature for some metals are given in Figures 5.2 and 5.3.

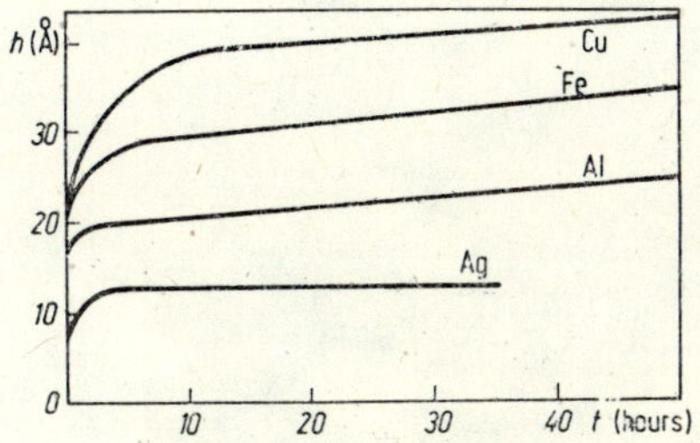

Figure 5.2 Thickness of the oxide layer formed at the temperature of 25°C for several metals [105].

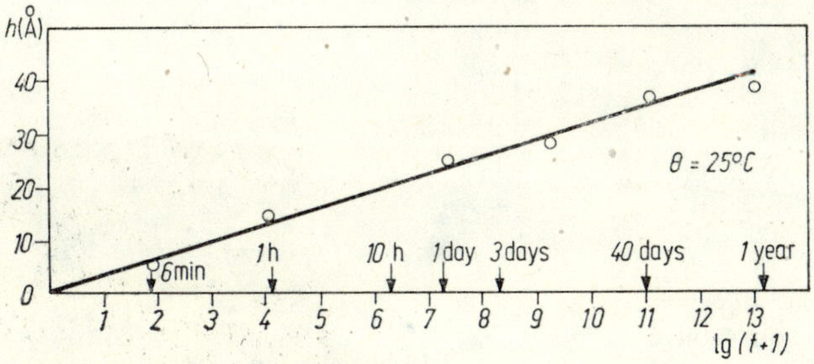

Figure 5.3 Thickness of the oxide layer formed on an iron or mild steel surface at the temperature of 25°C [105] (exposure time in minutes).

5.2 Dry friction and wear of metallic materials

Friction occurring in the absence of a lubricant is termed dry friction. The surfaces of the friction pair are however covered either by an absorbed layer of gases or vapours or by reaction products of the surface material and the gaseous atmosphere. In the case of friction in air the surfaces will inevitably be covered by an oxide layer. The formation and preservation of the oxide layer on the surfaces depend on the running conditions of the pair.

5.2.1 Wear of metallic surfaces. Qualitative aspects

The wear of metallic surfaces in air has several characteristics determined by the nature of the materials of the friction pair and the parameters of the process. To obtain valid results, the surfaces should be prepared so that there is no possibility of contamination with organic layers. Thus, these surfaces will always run under carefully controlled conditions from the beginning [92].

This chapter is devoted to some characteristic aspects of the process of wear of clean metallic surfaces in air, and particularly to the effect of the nature of the surfaces and load on wear. The apparatus and pair used were shown in Figure 4.2 and described in Chapter 4.1.

Figure 5.4 Bronze bush. Friction pair: bronze on bronze:
load: $P = 2$ daN; sliding speed: $V = 0.4$ cm/s; operating time: 2′.

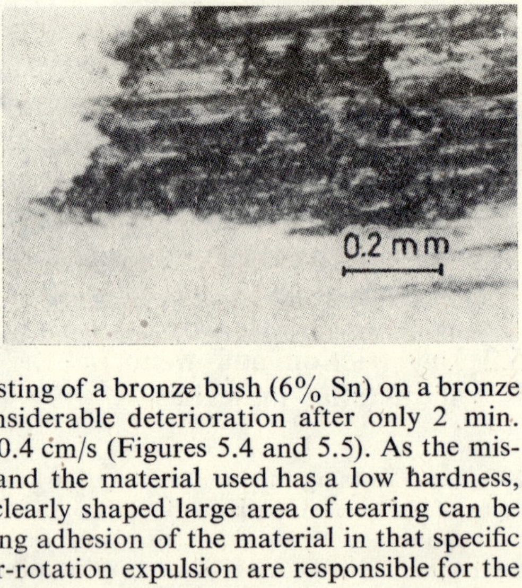

Figure 5.5 Bronze plate. Friction pair: bronze on bronze:
$P = 2$ daN; $V = 0.4$ cm/s; operating time: 2′.

In the case of a friction pair consisting of a bronze bush (6% Sn) on a bronze plate, the frictional surfaces suffer considerable deterioration after only 2 min. running time at a low sliding speed of 0.4 cm/s (Figures 5.4 and 5.5). As the miscibility of the materials is maximum and the material used has a low hardness, intense adhesive-type wear occurs. A clearly shaped large area of tearing can be observed on the bush surface. The strong adhesion of the material in that specific area to the fixed surface and its counter-rotation expulsion are responsible for the formation of a crater.

If in the pair described above the bronze plate is replaced by a quenched high speed steel plate, the bronze bush suffers considerable wear (Figures 5.6 and 5.7). As the bronze used consisted of a single phase crystallized in the cubic system the adhesion and wear of the bronze surface were severe. However, it may be noted that although the bushes in Figures 5.5 and 5.7 had the same running time of 2 min. with the latter the wear of the bronze bush was less severe and more uniformly distributed. This could be attributed to the reduced tendency to adhesion and a more uniform load distribution. The steel plate was slightly worn. Significant bronze transfer is evident on the surface.

Examination of the wear scar after only 2 min. running time allows the determination of the time when both worn and unworn areas coexist on the surface. Thus, the formation of "pittings" could be better observed. If the running time of the pair is increased, the whole surface deteriorates. In addition to specific forms of adhesive-type wear, abrasive-type wear scars also occur.

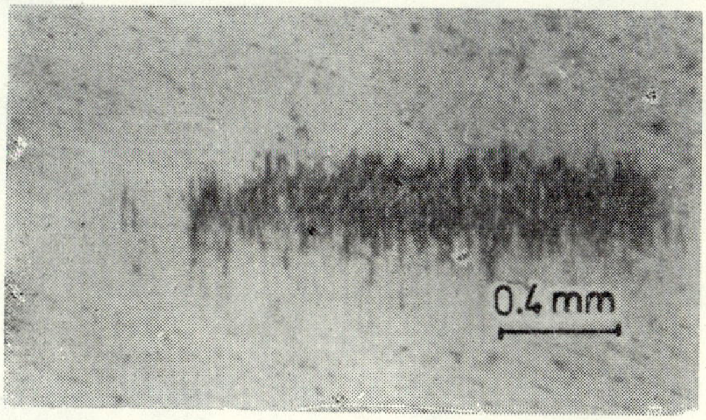

Figure 5.6 High speed steel plate. Friction pair: bronze on high speed steel: $P = 5$ daN; $V = 0.4$ cm/s; operating time: 2′.

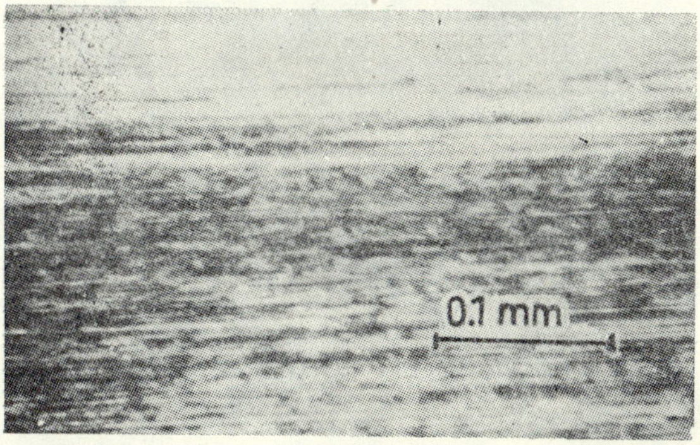

Figure 5.7 Bronze bush. Friction pair: bronze on high speed steel: $P = 5$ daN; $V = 0.4$ cm/s; operating time: 2′.

Increasing the running time to 60 min. causes the processes reported after running for 2 min. to be extended over the whole surface. As distinguished from the test lasting 2 min., the wear scar is much wider and completely covered with transferred bronze. After a short running time, the subsequent contact takes place between two bronze surfaces. Figure 5.8 shows the worn bush surface indicating the generation of a new wear particle and wear scars by a cutting process with hard bronze particles. In this case the bronze particle which acted as a cutting tool separated eventually from the steel surface and became attached to the bush. During one of the subsequent passages over the loaded area this particle will separate forming an independent wear particle.

At the speed of 60 cm/s with the same bronze on steel pair it is found that although the transfer of bronze to the steel surface still takes place, several additional events also occur, namely a reduction in transfer with smaller wear particles and wear scars. Figure 5.9 shows the agglomeration of wear particles in the entrance region of the loaded area. The fact that a considerable number of abrasive wear scars are produced on the steel plate is important (Figure 5.10). An enlarged clearer image of this wear feature was obtained by examining the scar with an

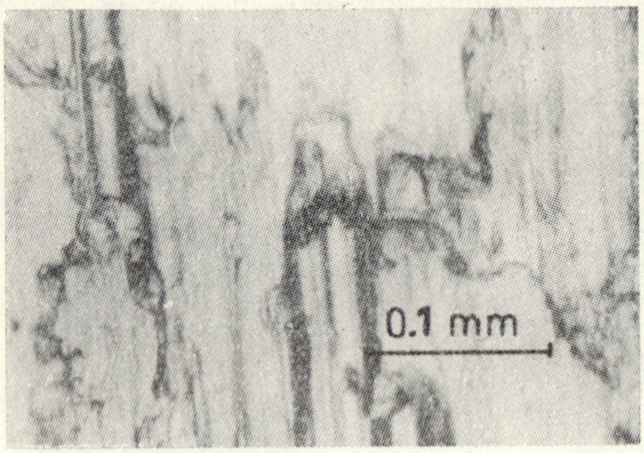

Figure 5.8 Bronze bush. Friction pair: bronze on high speed steel: $P = 10$ daN; $V = 0.4$ cm/s; operating time: 60′.

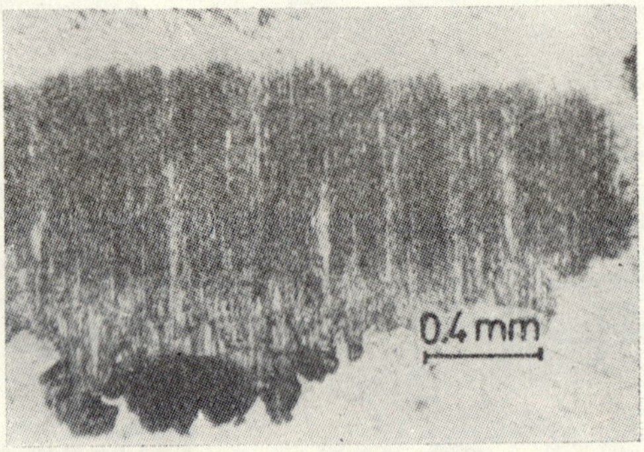

Figure 5.9 High speed steel plate. Friction pair: bronze on high speed steel: $P = 5$ daN; $V = 60$ cm/s; operating time: 2′.

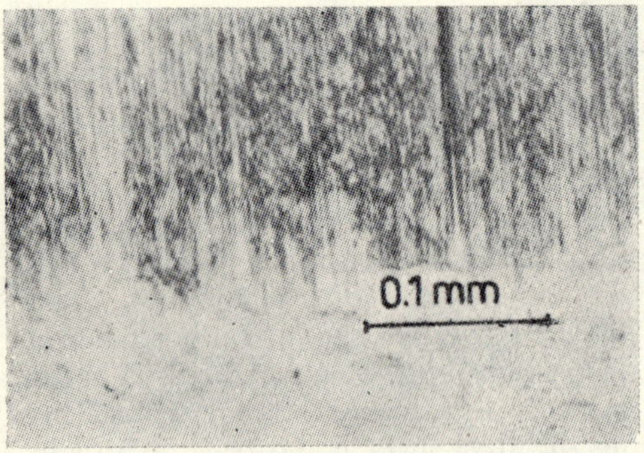

Figure 5.10 High speed steel plate. Friction pair: bronze on high speed steel: $P = 5$ daN; $V = 60$ cm/s; operating time: 2′.

electron microscope (Figure 5.11). Although the hardness of the bronze surface is only HB 130 kgf/mm² abrasive wear of the steel surface (of hardness HRC 65) still occurs. It is possible that this abrasive wear process occurs by means of fully cold work hardened particles or by steel particles separated from the plate and attached to the bronze surface.

The use of a friction pair in which both parts are of steel determines a considerable reduction in material transfer and formation of an oxide layer on the area of contact (Figures 5.12 and 5.13). There are two stages in this process of

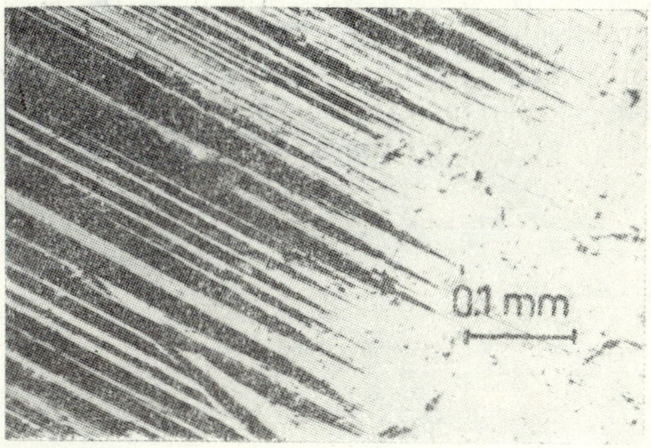

Figure 5.11 High speed steel plate. Friction pair: bronze on high speed steel: $P = 5$ daN; $V = 60$ cm/s; operating time: 2′. Image obtained by the emission electron microscope. Electron excitation: 30 keV.

Figure 5.12 High speed steel plate. Friction pair: high speed steel on high speed steel: $P = 10$ daN; $V = 0.4$ cm/s; operating time: 2′.

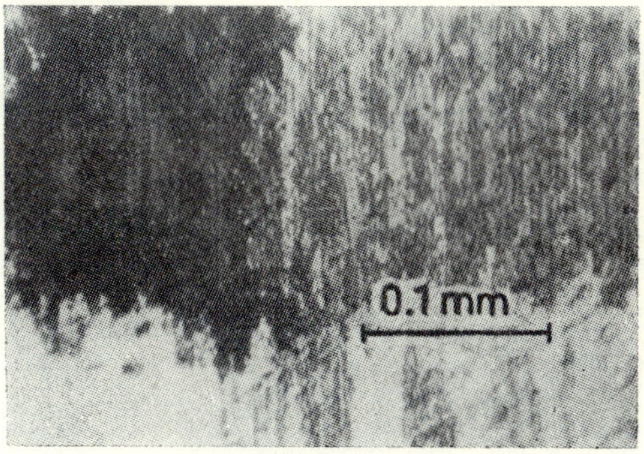

Figure 5.13 High speed steel plate. Friction pair: high speed steel on high speed steel: $P = 10$ daN; $V = 0.4$ cm/s; operating time: 60′.

oxidative wear: oxidation of the surfaces and formation of a thin adherent oxide layer followed by separation and agglomeration of this layer at the entrance to the loaded area. In this case, the oxidation of the surfaces plays the major role in the frictional process, contributing to the separation of the surfaces on the one hand, and to wear by abrasion on the other hand.

5.2.2 Wear of metallic surfaces. Quantitative aspects

As distinguished from the wear process in the presence of a lubricant which is characterised by a wide spread of the results, wear under dry friction conditions is reproducible with sufficient precision to allow the assessment of some quantitative relationships between the main parameters defining this type of wear. The scatter in results with lubricated wear is due to the considerable difference existing between the wear rates in various domains of mixed lubrication and to

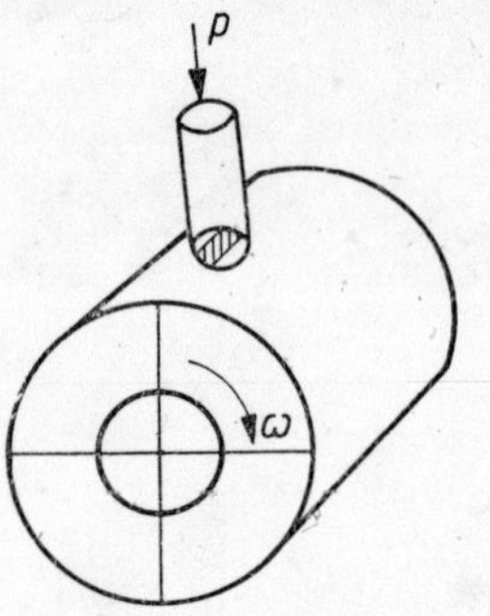

Figure 5.14 Friction pair for the study of dry wear.

the difficulty of keeping the lubrication conditions constant. An analysis of the poor reproducibility in lubricated wear test devices is given in Chapter 8.

Archard [8] used a simple pin on bush device to investigate wear under conditions of dry friction. With this device the load and sliding speed can be

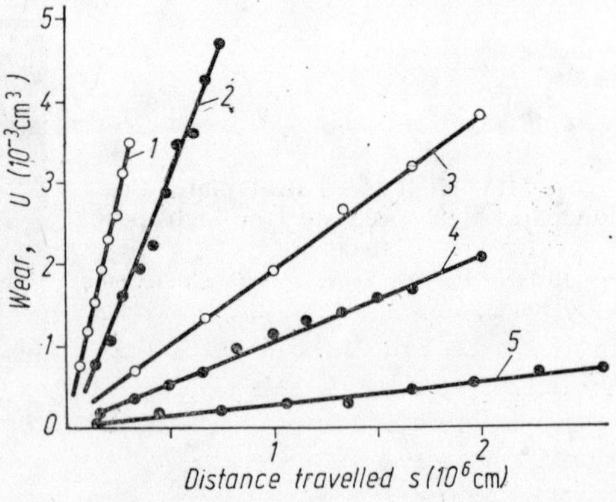

Figure 5.15 Variation of pin wear with distance travelled (after [8]): $V = 180$ cm/s; *1*, teflon on quenched tool steel, $P = 150$ g; *2*, brass (40% Zn) on quenched tool steel, $P = 120$ g; *3*, stainless steel on sintered tungsten carbide, $P = 1\,200$ g; *4*, bronze with beryllium on quenched tool steel, $P = 1000$ g; *5*, sintered tungsten carbide on mild steel, $P = 1000$ g.

varied within broad limits (Figure 5.14). For a certain load *P* and a sliding speed *V*, the pin wear *U*, determined either gravimetrically or by measurement of the wear scar size, depends linearly on the distance travelled *s* (Figure 5.15). The linear relationship always occurs after a specific time which is required to achieve a steady state on the frictional surface of the pair. For comparison, the same figure shows the wear of a plastics on metal pair. Curves similar to those in Figure 5.15 were obtained over a wide range of sliding speeds and loads. The linear portion of the curve $U = f(s)$ occurs before the pin surface conforms to the bush surface. Thus, the important conclusion drawn was that the wear rate, that is the slope of the lines plotted in Figure 5.15 is independent of the apparent contact area.

The dependence of wear U on load may also be deduced from some curves similar to those shown in Figure 5.15, but determined at different loads (Figure 5.16).

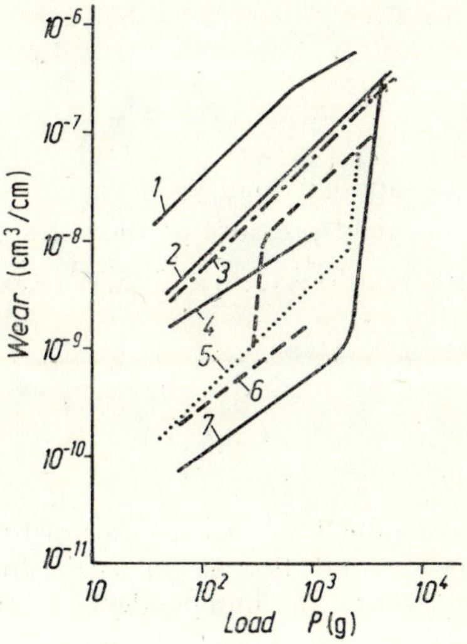

Figure 5.16 Variation of wear rate with load, $V = 180$ cm/s:
(after [8]): *1*, mild steel on mild steel; *2*, brass (40% Zn) on quenched tool steel; *3*, teflon on quenched tool steel; *4*, stainless steel on stainless steel; *5*, brass (30% Zn) on quenched tool steel; *6*, quenched tool steel on quenched tool steel; *7*, stainless steel on quenched tool steel.

Generally the wear intensity $U : s$ depends linearly on the load, each material being characterized by a specific wear resistance. In some particular instances this may not be linear or may even exhibit discontinuities (such is the case of material 5 or 7). To account for this behaviour the processes which take place in the frictional area under the influence of the running conditions should be considered.

As shown above, the process of dry friction is controlled by the nature of the surfaces which depend on the nature of the material of the friction pair, the environment and the operating conditions such as sliding speed, load and surface temperature. In the linear portion of the curves plotted in Figure 5.16 the dynamic equilibrium between surface wear and formation of the surface layer is established. Under a certain load or at a certain sliding speed another state of the surface becomes stable due to an increase in temperature and deformation which results in a change in the wear rate.

To explain wear, that is the process of separation of material from the surfaces of the friction pair as a consequence of their relative displacement, Archard proposed a mechanism based on asperity interaction. The probability of formation of a wear particle during a contact between two asperities is K_1. As the probability K_1 is small and the average number of contacts after the sliding distance s has been travelled is the same for all the asperities of the real area of contact A, the worn volume of material can be written as

$$U = K s A. \tag{5.6}$$

In relation (5.6) the constant K may be interpreted as the fraction of the contact surface which produces free wear. If a plastic contact is assumed to exist between the surfaces, then the real area of contact A is

$$A = \frac{P}{\sigma'_c}, \tag{5.7}$$

where

P is applied load,

σ'_c — yield pressure of the softer material.

From relations (5.6) and (5.7) an expression for the worn material is obtained

$$U = K \frac{sP}{\sigma'_c} \tag{5.8}$$

The model leads to a relationship which is in good agreement with the main experimental data as far as the linear dependence of wear on the distance s and load P and the independence of the apparent area of contact.

The constant K, which may be defined only under constant operating conditions, characterizes the materials of the friction pair under specific conditions existing on the frictional surface. When these conditions change due to increase in load, sliding speed or temperature, constant K changes. The main difficulty associated with wear calculation is in the determination of this constant and of its dependence on the operating conditions. Owing to the complexity nature of the dry friction process, the value of K can only be determined experimentally. Table 5.3 gives the wear intensity, hardness and constant K for several pairs of materials. It may be noted that the constant K has small values spread over a wide range. The wear constant tends to decrease with increase in hardness. This is valid especially for related materials. By using a different type of material smaller values of K may be obtained although the hardness is lower. Such is the case for rolled bakelite as compared with quenched tool steel (Table 5.3).

Relation (5.8) has been derived by assuming a plastic contact between the parts of the friction pair. The small value of the wear constant K, i.e. the large number of contacts required to form a wear particle, makes the plastic contact assumption questionable. A repeated plastic deformation of the contacting asperities under constant load cannot be conceived. For this reason relation (5.8) is also used in the form $U = K'sP$, where K' incorporates all the antiwear and mechanical properties of the material.

Table 5.3

Values of *K* for several materials (after [8])

Combination of materials	Wear intensity (10^{-10}cm^3/cm)	Hardness (10^6 g/cm^2)	Wear constant *K*
Soft steel on soft steel	1570	18.6	$7 \cdot 10^{-3}$
Brass (40% Zn)	240	9.5	$6 \cdot 10^{-4}$
Teflon	200	0.5	$2.5 \cdot 10^{-5}$
Brass (30% Zn)	100	6.8	$1.7 \cdot 10^{-4}$
Cast bakelite	12	2.5	$7.5 \cdot 10^{-6}$
Silver steel	7.5	32	$6 \cdot 10^{-5}$
Quenched tool steel	6.0	85	$1.3 \cdot 10^{-4}$
Stainless steel	2.7	25	$1.7 \cdot 10^{-5}$
Rolled bakelite	1.8	3.3	$1.5 \cdot 10^{-6}$
Sintered wolfram carbide on soft steel	0.9	18.5	$4 \cdot 10^{-6}$
Sintered wolfram carbide on wolfram carbide	0.03	130	$1 \cdot 10^{-6}$

The measurements were carried out using the friction pair shown in Figure 5.14 under a load $P = 400$ g and a sliding speed $V = 180$ cm/s. Except for the mentioned cases, the bush was made of quenched tool steel.

5.2.3 The wear mechanism and the role of the oxide layer

The formation of free wear particles is the final result of a complex process of interaction between the two surfaces of the friction pair. A useful technique for examining the various stages of the wear process is to use radioactive tracers [102, 9]. With the apparatus shown in Figure 5.14 a radioactive pin may be used instead of the conventional pin.

The high sensitivity of this technique allows both the total wear and the material transferred to the bush surface to be observed. An important achievement of the radioisotope technique is that it establishes the transfer of material from the pin to the bush. The transfer process is closely related to the wear of the friction pair as it has been established that the steady-state wear regime is reached only after the quantity of transferred metal has reached a constant value (Figure 5.17).

The dynamics of the transfer process may be further examined by replacing the radioactive pin with a previously run-in non-radioactive pin of the same material after the friction pairs have reached the steady-state regime. A reduction in the bush activity indicates the existence of a dynamic equilibrium of the transferred layer. In the steady-state regime the amount of material detached from the

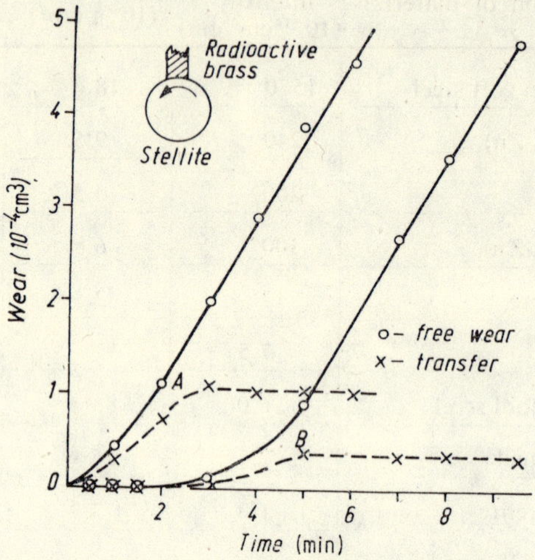

Figure 5.17 Variation of wear and material transfer with time [102]:
A, unlubricated, $P = 5$ kg; *B*, lubricated with cetane, $P = 22.5$ kg.

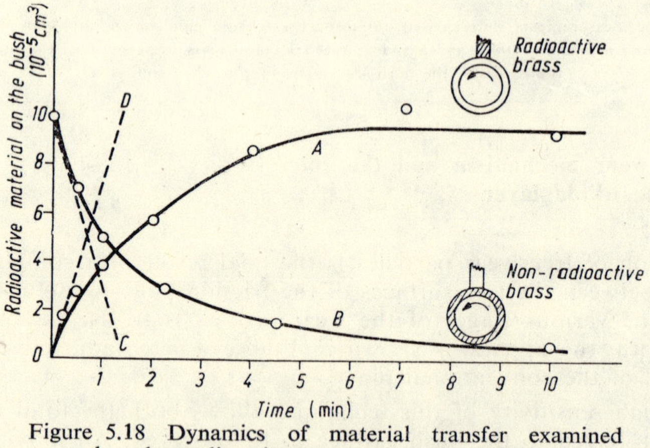

Figure 5.18 Dynamics of material transfer examined by the radioactive tracers technique [102]:
A, formation of radioactive transferred layer; *B*, removal of radioactive layer.

bush surface as free wear particles is replaced by an equal amount of material transferred from the surface of the pin. A certain radioactivity on the surface of the non-radioactive pin has also been noted (Figure 5.18). The origin slope of the "time-volume of transferred material curves" is equal to the wear rate of the pin. This indicates that the entire amount of worn material comes from transferred material.

Examination of the wear products, wear particle size and surface condition leads to the conclusion that two wear regimes exist depending on the load on the friction pair. According to the characteristic wear intensity these are termed mild and severe wear regimes respectively.

The mild wear regime which occurs under moderate loads is characterized by the small size of the wear particles (~ 100 Å) and the low roughness of the surfaces of the friction pair. Examination of wear debris has shown that for metallic pairs oxides are the main compounds. Thus, for ferrous alloy pairs the presence of Fe_3O_4, Fe_2O_3 and α-Fe was established by X-ray and electron diffraction. The proportion of the various compounds is determined by the magnitude of the load and the composition of the environment [145]. Considering these facts the following mechanism may be proposed for the process of mild wear of metallic friction pairs:

— transfer from the fixed part to the moving part,

— oxidation of the transferred material and of the fixed surface,

— transfer of part of the transferred material on the moving part back to the fixed part,

— separation of free wear particles from the transferred layers.

The above processes take place simultaneously at variable rates until the steady-state regime is reached.

The mild wear of unlubricated metallic pairs under a normal atmosphere may be considered to be a process of moderate abrasion of several oxide layers by oxide particles coming from these layers. In the case of metallic friction pairs the oxide layer determines the moderate character of the wear process. If the oxide layer cannot form or forms too slowly, wear of the metallic friction pairs increases.

The role played by the oxide layer in the friction of unlubricated metallic friction pairs may also be established by measurement of the coefficient of friction [94]. A bronze on steel pair with freshly processed surfaces has a coefficient of friction of $\mu = 0.55$; following oxidation for 24 hours this coefficient is reduced to $\mu = 0.3$ (Figure 5.19).

The formation of an oxide layer on the surfaces of the friction pair was also established under lubricated conditions [72, 96]. In this instance oxidation of the surface occurs due to oxygen dissolved in the lubricant. Generally, the presence of an oxide layer increases the adsorption capacity of the surface which leads to an improvement in the load carrying capacity of the friction pair under conditions of boundary and mixed lubrication.

If the load increases, the formation rate becomes smaller than the wear rate and the protective oxide layer is removed. Under such conditions the friction pairs are subjected to a severe wear regime which is characterized by a high wear rate, great surface roughness and wear particles of a much larger mean diameter (0.01 — 0.1 mm). The main mode of interaction of the surfaces is adhesion in the area of contact. A concentration of normal stresses as a consequence of the increase in the wear particle diameter is also noted. The severe wear regime is considered to occur when the wear particle diameter corresponds to that of the real area of contact. It has also been found for several alloys that the transition from mild to severe wear occurs when the maximum shear stress in the contacting regions reaches one-sixth of the hardness of the softer material [195].

The severe wear process initially occurs at discrete areas on the surface (Figure 7.9). These areas develop and the complete surfaces of the friction pair

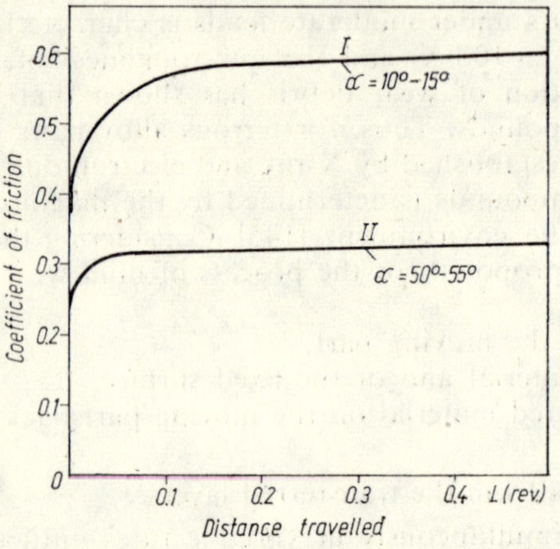

Figure 5.19 Effect of the oxide layer on the coefficient of friction for a pair of bronze on steel. $V = 4 \cdot 10^{-3}$ cm/s; I, freshly prepared surfaces; II, oxidized surfaces (α is the surface contact angle with respect to glycerine).

become involved. Development of the severe wear regime is also attributed to a cumulative process. The occurrence of the severe wear regime over certain areas of the surface causes an increase in the coefficient of friction which leads to a rise in surface temperature.

Because of its intensity, severe wear is a destructive process which rapidly causes machine component failure. An example of severe wear occurs in the seizure of lubricated friction pairs.

5.2.4 **Rate of oxidative wear**

Assuming several simplifications imposed by the complexity of oxidation and wear processes, a relationship for oxidative wear rate may be derived as a function of load and material characteristics [145].

It is assumed that:

— the bulk temperature of the body is important in the oxidation process as for most friction pairs the duration of this temperature is much longer than that of the flash temperature,

— when the oxide layer reaches a certain thickness it separates to form a wear particle,

— the processes of oxidation and wear occur on the peaks of asperities; the wear particles are small in size.

Consider a logarithmic law for the oxidation rate (valid for steel below 200°C)

$$h = C \ln\left(\frac{t}{B} + 1\right), \tag{5.9}$$

where

h is thickness of the oxide layer,
C – coefficient function of temperature,
B – material constant,
t – time.

As $B \simeq 10^9\, s$ relation (5.9) can be simplified as

$$h = \frac{C}{B} t. \tag{5.10}$$

If for the coefficient C an expression similar to relation (5.3) is used the thickness h may be written as

$$h = \frac{C_0}{B} t\, \mathrm{e}^{-\frac{Q}{R\theta_V}}, \tag{5.11}$$

where θ_V is the bulk temperature; Q is the activation heat and R is a constant for perfect gases.

Assume that h_c is the thickness at which the oxide particle separates, corresponding to a time t_c. The volume of worn material after a time t_c is

$$V_c = A\, h_c, \tag{5.12}$$

where

$A = \frac{P}{H}$ is the real area of contact for the plastic contact,
P – load,
H – hardness of the softer material.

It follows from relations (5.11) and (5.12)

$$\frac{h_c A}{t_c} = \frac{P C_0}{B H} \mathrm{e}^{-\frac{Q}{R\theta_V}}. \tag{5.13}$$

Thus, an expression is found for the mean wear rate, $h_c A/t_c$, which is a function of the material properties H, C_0, B, Q and the operating conditions of the friction pair, namely the load P, the sliding speed and the coefficient of friction. This relationship does not contain the thickness h_c at which the oxide layer is considered to separate.

5.2.5 Size of the wear particle

Rabinowicz proposed a method for the calculation of wear particle size using the energy of adhesion between the surfaces of the friction pair [146]. If plastic deformation occurs at the time of contact between a spherical asperity of the surface of lower hardness and the flat surface of the counterpart having a higher hardness, then the elastic deformation over the contact area reaches the value ε_{max}. A tangential deformation $\nu\varepsilon_{max}$ will simultaneously occur, where ν is Poisson's ratio. Tangential elastic deformation may still be present even if the load is no longer applied due to the friction between the particle and the surface. The elastic deformation energy of an asperity of diameter d is

$$E_v = \frac{1}{2}\nu^2\varepsilon_{max}^2 E\frac{\pi d^3}{12}. \tag{5.14}$$

Considering that the whole surface of the spherical asperity is in contact with the harder surface, the surface energy of the particle is

$$E_s = E_{ab}\frac{\pi d^2}{4}, \tag{5.15}$$

where E_{ab} is the energy of adhesion between surfaces a and b.

For a wear particle of diameter d to form it is necessary that

$$E_v \geqslant E_s. \tag{5.16}$$

Condition (5.16) together with relations (5.14) and (5.15) give the wear particle diameter. For common unquenched steel, assuming $\varepsilon_{max}E = \frac{H}{3}$, $\nu^2 \simeq 0.1$ and $\varepsilon_{max} = 3.10^{-3}$, the particle diameter will be approximately

$$d \geqslant 60\,000\frac{E_{ab}}{H}, \tag{5.17}$$

where H is the hardness of the softer material.

Table 5.4

Mean diameter of the wear particle for several lubricants. Friction pair: copper on copper; Load: 1.100 g [147]

Lubricant	Mean diameter of the wear particle (μm)
N_2	480
He_2	380
CO_2	300
Dry air	224
O_2	201
Normal air	177
Wet air	144
Cetane	12
Siliconic oil	9.5
Palmitic acid in cetane	8

The validity of relation (5.17) has been confirmed under certain conditions both for dry and lubricated friction [147]. It is known that adsorption of lubricant on the frictional surfaces produces a reduction in their surface energy. The more effective the lubricant under boundary lubrication conditions, the greater the reduction in surface energy. According to relation (5.17) the increase in the effectiveness of a lubricant causes a reduction in the mean diameter of the wear particle, which was confirmed experimentally (Table 5.4).

5.3 Friction in special environmental conditions

Most machines and mechanisms operate in a normal atmosphere over a relatively narrow temperature range. The whole development of these mechanisms, namely the design methods, the materials and lubricants selected take into consideration these conditions which are considered to be constant. The development of new technologies or research fields especially in the last decades has imposed the necessity of devising mechanisms capable of operating under different, unusual environmental conditions. For instance, aerospace engineering has brought new problems such as high vacuum and a high degree of reliability. In nuclear energy fields and especially in nuclear reactors cooled with He_2 or CO_2 the operating conditions are affected by the absence of oxygen on the one hand and by the high temperature and ionized radiations on the other hand. Other examples of arduous conditions are the hot working of metals and the machining of special alloys where conventional lubrication techniques cannot meet the specific requirements.

Special operating conditions involve the following fields:

— operation in vacuum or in the presence of several gases other than air,

— operation under conditions of extremely high or extremely low temperatures,

— simultaneous presence of extreme temperatures and of a vacuum.

5.3.1 Operation in vacuum. Lubricants and materials used

Reduction in the atmospheric pressure has a twofold effect, namely:

(a) it changes the tribological characteristics of the metallic surfaces by reducing the rate or even suppressing surface oxidation,

(b) it changes the composition and structure of the lubricant, which affects lubricating ability.

The oxide layer plays an important role in the process of friction between metallic surfaces. For most commercial alloys, the presence of this layer ensures an ultimate protection of the surfaces against severe wear and seizure.

The severity of the frictional process in the absence of oxygen may be established also by measuring the coefficient of friction. Figure 5.20 shows the effect of prolonged vacuum on the coefficients of friction of a nickel on wolfram and a copper on copper friction pair. The coefficient of friction rises from a value of 0.5 which is characteristic of air to $4 \div 5$. By introducing a very small amount of oxygen the coefficient of friction decreases again to low values. The results plotted in Figure 5.20 were obtained for light loads and for a single passage [39].

Vacuum affects the lubricant by accelerating evaporation and by structural modifications. The lower the molecular weight, the greater the effect of pressure decrease or temperature rise on evaporation rate (Figure 5.21).

The different evaporation rates of the constituents of a lubricant, which in most cases are mixtures of several substances, will finally result in a selective separation of the constituents. The remaining fraction will be richer in constit-

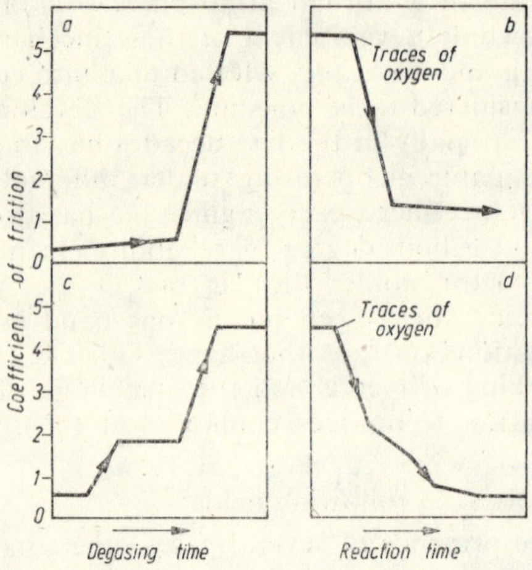

Figure 5.20 Effect of oxygen on the coefficient of friction of metallic pairs [39]:
a and *b*, nickel on wolfram; *c* and *d*, copper on copper.

uents of high molecular weight. The same process of selective evaporation also occurs in greases. If the fluid component of the grease is volatile under the conditions of pressure and temperature used, enriching of the grease in thickening agent occurs with time.

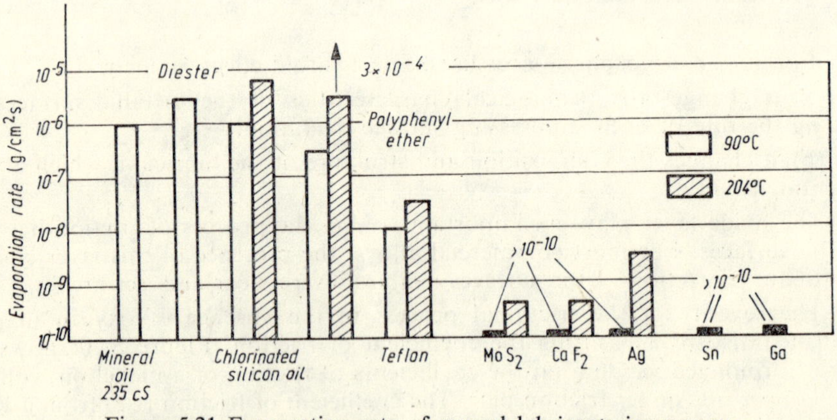

Figure 5.21 Evaporation rate of several lubricants in vacuum ($10^{-6} \div 10^{-7}$ torr) [114].

Prolonged operation under conditions of low pressure may also cause a dissociation of the lubricant molecules and hence a reduction in the mean molecular weight of the lubricant. This process was established in the lubrication of rotary vacuum pumps by examining the released vapours by mass spectroscopy.

To devise a mechanism with satisfactory operation in vacuum the above effects should be minimized by adequate design incorporating suitable lubricants and materials.

Up to pressures of $10^{-6} \div 10^{-7}$ torr and temperatures of $100 \div 120$°C fluid lubricants or greases may be used. The lubricant used or that incorporated in the grease should be of a sufficiently high molecular weight to minimize the evaporation rate. If the total lubricant volume is limited, the fraction with high molecular weight should be representative. Also, the lubricant should contain additives which are capable either of reacting with the metallic surface or of having reactive radicals incorporated in their molecule. Thus the reaction layer which forms takes over the role of the oxide layer.

The evaporation rate of the lubricant may also be considerably reduced by a correct design solution which should increase the hydraulic resistance in the vapour circuit to a maximum.

Conventional alloys possess their well-known tribological characteristics only if they are covered with an oxide layer. The absence of an oxide layer can be compensated for by using some non-metallic inclusions which are entrained to the pair surface in operation. Inclusions of oxides and sulphides gave the best results. Alloys with non-metallic inclusions, which are normally avoided for high quality parts because of their low fatigue strength, behave better in vacuum. The reduction in fatigue life is compensated for by an increase in the resistance to adhesion wear.

Another solution is to use Cu — Pb-type alloys in which a hard phase taking over the role of the support and a soft phase playing the part of the lubricant coexist. Good results have been reported for Ag — Ni, Sn — Ni and Si — Ni alloys. In the selection of the metals to be used for bearing surfaces their crystalline structure plays an important role. It was found that the metals with a hexagonal lattice exhibit a much lower coefficient of friction wear rate and seizure tendency than the metals with a face-centered cube lattice. Figure 5.22 gives the coefficient of friction for some metals with a hexagonal lattice.

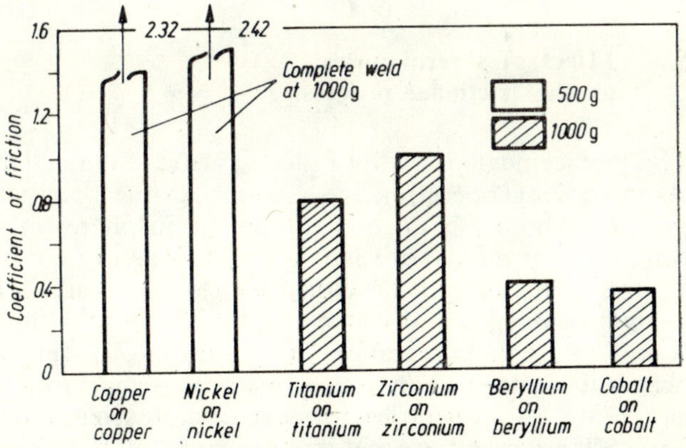

Figure 5.22 Coefficient of friction for several metals with cubic (Cu, Ni) and hexagonal (Ti, Zr, Be, Co) lattices [44].

The lanthanides too belong to the class of metals with a hexagonal lattice. It is seen that the coefficient of friction of these elements decreases as the length of the axis c of the elementary cell increases (Figure 5.23.). With the absence of any lubricant the coefficient of friction is approximately 0.2 for neodymium and approximately 0.4 for lanthanum and samarium, i.e. even lower than the coefficient of friction of steel in air. Because of their reduced mechanical strength, such metals may be used only in the form of thin layers.

For ultrahigh vacuum (of the order of 10^{-10} torr) fluid lubricants cannot be used because of their high evaporation rate and self-lubricating materials, solid lubricants, and soft metallic layers are used. The final choice of a lubricant should be based on relevant effective tests of the mechanism. The use of some

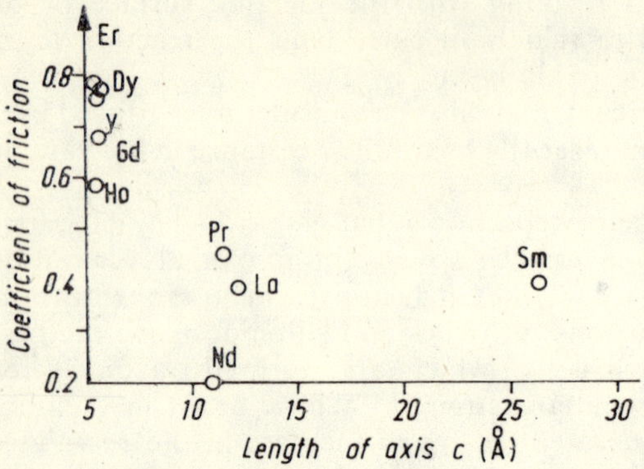

Figure 5.23 Variation of the coefficient of friction with the length of the axis c for several lanthanides [45]. Pressure: 10^{-9} torr; Load: 100 g; temperature: 25°C.

conventional alloys for a gear of Ø = 25 mm (transmitted moment: 800 gcm and $\omega = 1500$ r.p.m) resulted in a maximum operating time of only several hundred hours in a vacuum of 10^{-10} torr. Working surface of the gear was covered with a thin layer of Pb deposited by condensation or with MoS_2 with Pb as a binder. A considerably increased operating time was obtained with plastic gears reinforced with carbon fibres with a stainless steel pinion. Such a combination led to a gear life of over 3600 hours under the conditions described [82].

5.3.2 **Effect of several gases on the frictional process**

The replacement of air by other gases results partially in effects similar to thoss characterizing operation in vacuum. As the reactivity of metals with oxygen ie extremely high, an unconventional atmosphere only effects performance if the content of oxygen and water is very low as in high purity gases.

Summers-Smith described a case of seizure in an azote compressor with a content of oxygen and water less than 2 p.p.m. although normal fluid lubrication was used [165]. Lubrication of the piston-cylinder assembly for both stages was performed with oil from a tank which was constantly washed with the compressed gas. With this operation three cases of seizure occurred after approximately only 500 hours service before the cause of failure was determined. No further problems occurred in service when the cylinders were lubricated with oil which was in contact with the atmosphere.

Another instance indicating the importance of a gaseous environment is the pitting failure of ball bearings in pure helium [100]. Experimental life testing of the bearings which were to be used inside a nuclear reactor showed that in the presence of pure helium (a cooling agent in the reactor) at an operating temperature from 100°C to 200°C the operating time was reduced to 140 hours. The contamination of helium with oxygen and water vapour resulted in an increase in bearing life of approximately 500% (Table 5.5). For technological reasons an increase in oxygen and water content in helium was not permissible so an oxidizing agent was added to the grease used. A mixture of Na nitrite

Table 5.5

Effect of increasing the oxygen and water content of helium on bearing life, determined experimentally [100]. Lubricant: a grease with silica gel as thickening agent

Concentration		Number of tests	Average time to failure (hours)
Oxygen, p.p.m	Water, p.p.m		
1	2	9	140
11	2	2	180
27	2	3	304
51	2	3	376
11	7	1	432
51	11	1	500

and Na stearate in weight concentration of 5% each was added and the operating time of the bearings was increased to 3000 hours.

With certain gases having a high lubricant solubility, difficulties may occur due to a reduction in lubricant viscosity as a consequence of the gas dissolution. Table 5.6 shows the effect of several dissolved gases on lubricant viscosity. As solubility increases with pressure, this phenomenon becomes

Table 5.6

Effect of several dissolved gases on viscosity (saturated solution) [26]

Gas	Oil	Temperature, °C	Final viscosity: initial viscosity
He	diester	95	0.94
He	diester	24	1.02
He	mineral oil	95	0.95
He	mineral oil	25	0.99
CO_2	mineral oil	27	0.31
Ar	mineral oil	25	0.64

important for installations such as compressors which operate under pressure. To compensate for the extremely high solubility of several gases such as carbon dioxide, argon or hydrocarbons in mineral oils, compressors operating with such gases are lubricated with an oil which is more viscous than that used when the compressor is operated with air. An alternative solution for the preservation of high viscosity is to use synthetic lubricants. Hydrocarbons have a low solubility in these lubricants as compared with conventional lubricants.

6 Layers formed from solution

6.1 Adsorption in the presence of a lubricant

As seen in Chapter 4.1, the main drawback of homogeneous thin layers is their limited endurance. The wear process of the layer is not compensated by a process of regeneration. Active substances, such as anti-wear and anti-seizure additives in solution form are often used to ensure a layer constantly present on the frictional surfaces. The adsorbed or reaction layer, although subject to wear, can be regenerated by the additive in the lubricant.

The presence of the lubricant which acts as a solvent for the respective additives affects the formation and behaviour of the layers in several ways. Generally, the layer adsorbed on the solid surface is not homogeneous, its composition incorporating molecules both of adsorbate and of lubricant. If a reaction layer forms, then the presence of the lubricant determines the concentration of the reactants and hence the reaction rate [1, 75, 130].

6.1.1 Composition of the adsorbed layer with dilute solutions

With dilute solutions the surface fraction occupied by the molecules of lubricant (solvent) and the molecules of additive (solute) can be determined. It is assumed that the solution is diluted and the adsorbed layer is monomolecular; in this instance the layer behaves as an ideal solution. The constant of the adsorption process of the solvent-solute system is

$$K = \frac{N_2 a_1}{N_1 a_2}, \tag{6.1}$$

where

a_1 and a_2 represent the activity of the solvent (1) and the solute (2) respectively,

N_1 and N_2 are the surface concentrations expressed in the form of molar fraction for the two types of molecules.

For dilute solutions we may take a_1 — constant and $\frac{K}{a_1} = b$. As $N_1 + N_2 = 1$, it follows that

$$N_2 = \frac{ba_2}{1 + ba_2} \cdot \tag{6.2}$$

As

$$n_2 = N_2 n,$$

where n is the total number of molecules adsorbed by 1 g of adsorbent, then the surface fraction θ occupied by the additive molecules is given by

$$\theta = \frac{n_2}{n} = \frac{ba_2}{1 + ba_2} \cdot \tag{6.3}$$

With concentrated solutions account should be taken of the component concentration variation because of the process of adsorption.

The constant b may be written as a function of the adsorption heat Q and the absolute temperature T as

$$b = b' \, e^{\frac{Q}{RT}} . \tag{6.4}$$

6.1.2 **Examination of adsorption by means of radioactive tracers**

The radioactive tracer technique is useful for the determination of adsorption isotherms because of its sensitivity and selectivity. By introducing a radioactive atom into the additive molecule the concentration of the additive in the solution can be controlled before and after contact with the absorbent. It is assumed, however, that the mass difference between the radioactive and non-radioactive atoms does not affect this process, which is true in most cases, that is

$$\left(\frac{G_1}{G_2}\right)_i = \left(\frac{G_1}{G_2}\right)_f , \tag{6.5}$$

where

G_1 is mass of the non-radioactive substance,
G_2 — mass of the same substance, labelled,
i, f — initial and final or before and after the process.

Application of the radioactive tracer technique is based on relation (6.5).

For a solution the above equation can be written after a simple transformation as

$$\frac{(C_T)_i}{(C_T)_f} = \frac{(C_2)_i}{(C_2)_f} , \tag{6.6}$$

where

C_2 is concentration of the radioactive substance,
C_T — total concentration of the solute.

The specific radioactivity of the solution Λ_{ss}, taking account of equation (6.6) is

$$\Lambda_{ss} = \text{ct.}\ C_T. \tag{6.7}$$

Hence, the solute concentration can be established by measuring the radioactivity of the solution.

If the radioactivity is measured over the adsorbent surface, for instance on the surface of a friction pair, then the counting rate R is given by

$$R = \varepsilon_T \Lambda_s G_T, \tag{6.8}$$

where

ε_T is total efficiency of the radiometric measuring system,
G_T — amount of adsorbed substance,
Λ_s — specific radioactivity of the adsorbed substance.

Determination of adsorption isotherms. To determine the adsorption isotherms it is necessary to measure the quantity of adsorbed substance at various concentrations and to relate this value to the quantity of adsorbent. The concentration is determined by measuring the radio activity of the solution. As in the study of additives, the solvent (lubricant) is a hydrocarbon, an installation with a liquid scintillator may be used for measuring this activity. This type of apparatus possesses a linear measuring characteristic and may provide directly the absolute radio activity of the specimen.

Consider the following quantities:

G is total quantity of solute,
G_a — adsorbed quantity of solute,
G_r — quantity of solute remaining in the solution,
V_T — total volume of the solution,
C — solute concentration,
G_{ad} — quantity of adsorbent,
A_d — efficiency of the adsorbent.

The relationships existing between these quantities are

$$A_d = \frac{G_a}{G_{ad}}, \tag{6.9}$$

$$G = G_a + G_r, \tag{6.10}$$

$$G_r = CV_T. \tag{6.11}$$

The required isotherm is the function

$$A_d = f(C). \tag{6.12}$$

Using equation (6.9) from (6.11) it follows that

$$A_d = \frac{G}{G_{ad}} - \frac{V_T}{G_{ad}} C. \tag{6.13}$$

As a rule, V_T = constant and G_{ad} = constant are taken. In this instance equation (6.13) produces a family of lines which are of slope V_T/G_{ad} with abscissa

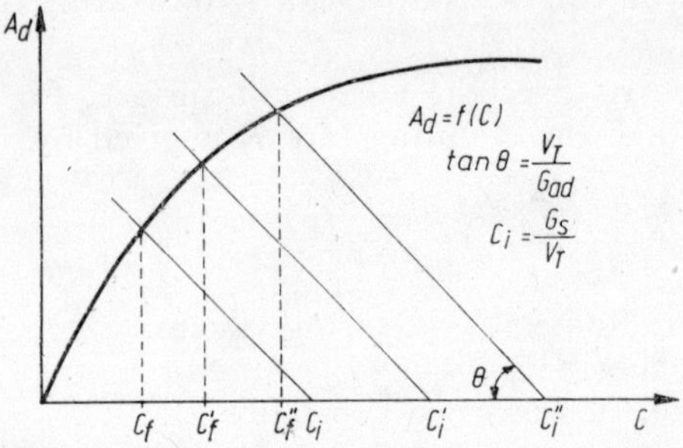

Figure 6.1 Determination of adsorption isotherm.

at the origin G/V_T which is the initial concentration (Figure 6.1). Determination of the isotherm is by measurement of the final radio activity of the solution for various initial concentrations.

If large specific radio activities of the solute are considered it is necessary to also add inactive solute to provide higher initial concentrations.

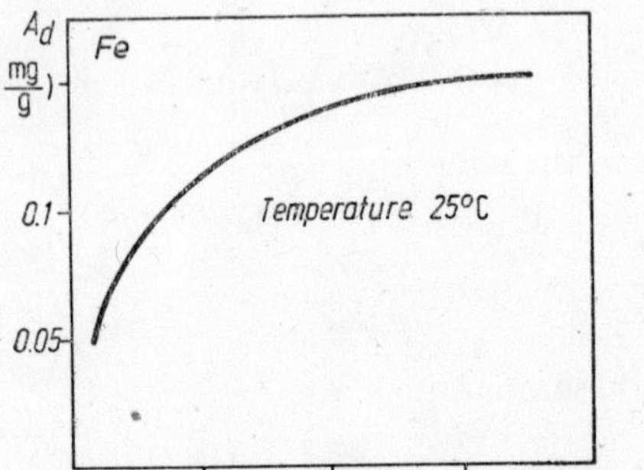

Figure 6.2 Adsorption isotherm for the aluminium oxide-dodecanoic acid system. Adsorbent: aluminium oxide; solute: dodecanoic acid; solvent: gasoline.

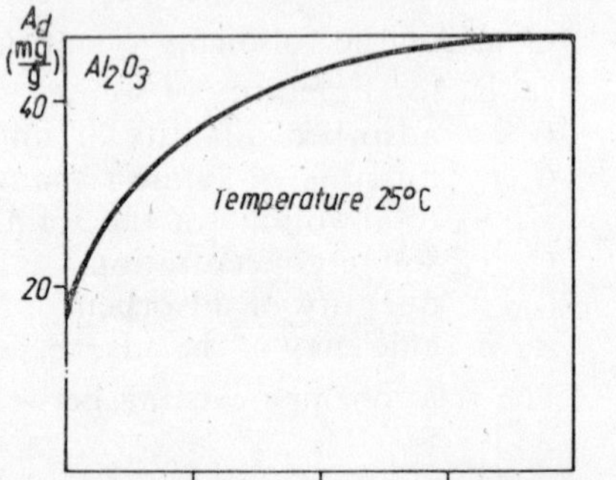

Figure 6.3 Adsorption isotherm for the iron-dodecanoic acid system. Adsorbent: iron powder; solute: dodecanoic acid; solvent: gasoline.

The technique described may be applied to solids in the form of powders which have a high specific surface. The technique allows the analysis of the interaction between the additive and the alloy at various temperatures. The same technique was used to determine the adsorption isotherms of dodecanoic acid from solution in gasoline on activated aluminium oxide and on iron powder (Figures 6.2 and 6.3) [92].

For the iron powder a low saturation concentration and a diminished adsorption capacity were noted. The saturation concentration is more important than the adsorption capacity which depends on the specific surface of the powder.

Saturation occurs after the entire surface of the solid body is covered with molecules of solute, so it would be expected that the effect on the frictional process should cease within the same concentration range. As shown in Chapter 6.3 the saturation concentrations determined by the two techniques are of the same order of magnitude.

The aluminium oxide, because of its high adsorption capacity, may be used to purify hydrocarbons from polar compounds. Certain tests on the effect of additives on friction require careful purification of the lubricants used. Any traces of polar compounds may affect the measurements. Purification is performed either by passing the liquid through a column of aluminium oxide or by repeated adsorption. As the efficiency of the adsorbent is known, the degree of purity achieved can be determined.

Examination of adsorption on surfaces. To determine directly the interaction between an additive and the surfaces of a friction pair, the radioactivity should be measured at the surface. Unlike the previous case, the quantity of adsorbed material is now determined. As the adsorbent surface is much less, the accuracy is lower. This technique is potentially suitable for the investigation of the effect of surface machining techniques and friction on the adsorption capacity. When measuring the surface radioactivity a front counter with a thin window or a gas circulation counter should be used with C^{14} or H^3 isotopes.

The activity measured on the surface is according to relation (6.8)

$$R = \varepsilon_T A_s G_T = k_2 G_T. \tag{6.14}$$

Determination of the constant k_2 requires a standard measured under the same conditions as the friction pair surface. Moreover, it should take into account the parameters used in adsorption on the surfaces of the friction pair. To achieve this standard the radio active substance deposited as the layer must fulfil thin layer condition to prevent errors due to self-adsorption [49].

The autoradiographic technique is also used to measure surface radioactivity as it has been found to be useful for the determination of the distribution of the radioactive element on the surface. This technique was used to study the stability of films of radioactive stearic acid and copper stearate on various surfaces (Table 6.1).

It is found that stearic acid and copper stearate are physically adsorbed and can be readily removed from non-reactive surfaces (platinum or mica). On the reactive copper surface the stearic acid forms in situ copper stearate which is

Table 6.1

Effect of surface on the stability of the films of stearic acid and copper stearate [113]

Surface	Property examined	Adsorbed material	
		stearic acid	copper stearate
Copper (probably in C_2O form)	(1) effect of heating in air stream at 81°C	not removed	not removed
	(2) mild abrasion	not removed	not removed
	(3) hot benzene	partially removed	partially removed
	(4) electron diffraction pattern	pattern corresponding to copper stearate	no pattern formed
Platinum	(1) effect of heating in air stream at 81°C	partially removed	partially removed
	(2) mild abrasion	not removed	not removed
	(3) hot benzene	totally removed	totally removed
	(4) electron diffraction pattern	pattern corresponding to the free acid	pattern corresponding to the copper stearate layer
Mica	(1) effect of heating in air stream at 81°C	totally removed	totally removed
	(2) mild abrasion	partially removed	partially removed
	(3) hot benzene	totally removed	totally removed

much more difficult to remove. The greater stability of the copper stearate layer formed in situ may be due to the formation of the monostearate in which the metal atom is partially bound to the oxide lattice.

6.2 Chemical interaction between the metallic surface and the lubricating medium

Physical adsorption and chemisorption first take place upon contact between the metallic surface and the lubricating medium. Usually, under the conditions existing on the friction surface, namely high temperature and surface activation by elastic or plastic deformation, the chemisorption process is continued by a reaction process. If the reaction layer satisfies certain conditions, the most important of which being the adhesion to the metallic surface, its presence will exert a beneficial effect on surface wear under conditions of mixed lubrication and increase the seizure load. The presence of a reaction layer on the active surfaces of the friction pair when additives are used has been established by various means such as radioactive isotopes, by physical or chemical techniques [37, 113, 153, 163].

Additive reaction at "hot" spots on the metallic surface may interfere with the process of boundary or mixed lubrication. The chemical reaction between the surface material and the additive prevents adhesion, as the reaction product has a much lower shearing resistance than the base metal. Thus, the junction will shear in the reaction layer and not within the metallic surfaces [39].

Kinetics of surface reactions. Several specific aspects of the kinetics of the reactions occurring during the lubrication of metallic surfaces have been investigated.

If it is assumed, for instance, that the rate of reaction between the chemisorbed layer and the surface is high, the total rate of the process is determined by the rate of diffusion of the additive molecules into the immobile boundary layer

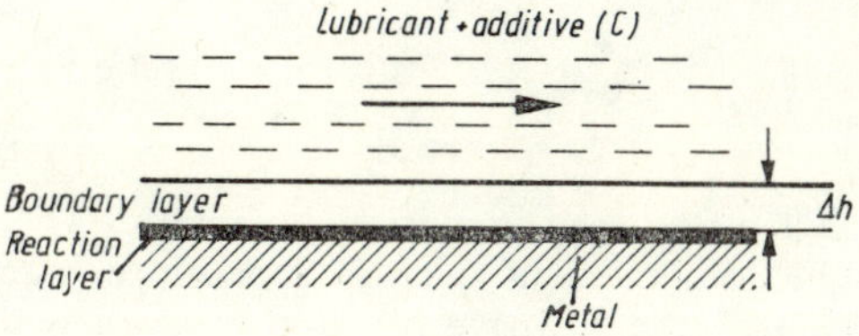

Figure 6.4 Diffusion of reactants through boundary layer and formation of reaction layer.

[3]. To reach the metallic surface the molecules of the reactants incorporated in the lubricant must diffuse through the boundary layer adhering to the surface (Figure 6.4). Reaction between the additive and the surface generally takes place only in the presence of water and oxygen. Under normal conditions, in the presence of air, mineral oils dissolve approximately 2% vol. oxygen or 0.0009 mol/l and approximately 0.0008 mol/l water. Consequently, at concentrations of the additive less than 0.001 mol/l, the reaction rate is determined by the additive concentration and for concentrations of the additive higher than 0.002 mol/l the reaction rate is determined by the concentration of oxygen and water.

The rate of diffusion through the boundary layer towards the frictional surfaces in a well-mixed lubricant is

$$\frac{dn}{dt} = \frac{D \cdot C \cdot N}{\Delta h \cdot 1000}\left(1 - \frac{n}{n_\infty}\right), \tag{6.15}$$

where

n is number of molecules (controlling the reaction) existing on the surface,

n_∞ — maximum number of molecules that may exist on the surface,

t — time,

D — coefficient of diffusion,

C — concentration of the reactant controlling the reaction,

Δh — thickness of the immobile boundary layer,

N — Avogadro's number.

The coefficient of diffusion for a certain oil and additive depends on the absolute temperature T, the lubricant viscosity and a constant b

$$D = \frac{bT}{\eta}. \tag{6.16}$$

Equation (6.15) by integration and replacing of the coefficient of diffusion by relation (6.16) gives

$$\frac{n}{n_\infty} = 1 - e^{-\frac{CTtNb}{\eta \cdot 1000 \cdot \Delta h n_\infty}}. \tag{6.17}$$

The diffusion time t may be expressed as a function of the sliding speed and a surface parameter. If it is assumed that the reaction layer is removed when contact occurs between two asperities and that the mean distance between two asperities is l, then the available reaction (diffusion) time is $t = l/V$, where V is the sliding speed. Replacing t in equation (6.17) it follows that

$$\frac{n}{n_\infty} = 1 - e^{-\frac{CTd}{\eta \cdot V}}, \tag{6.18}$$

where d is a constant including N, b, l and n_∞.

The coefficient of friction μ or more exactly the reduction in the coefficient of friction of the additive oil as compared with the straight oil [3] indicates the extent to which the surface is covered with a reaction layer

$$\frac{(\Delta\mu)_{V,T}}{(\Delta\mu)_{V=0,T}} = \left(\frac{n}{n_\infty}\right)_{V,T}, \tag{6.19}$$

where $\Delta\mu = \mu$ (oil + additive) $- \mu$ (oil) at the velocity V and temperature T.

By substituting equation (6.19) into equation (6.18) an equation may be obtained between the coefficient of friction and the principal factors influencing the frictional process, namely temperature, sliding speed and lubricant viscosity. As equation (6.19) is based on a simple model its validity is limited.

The kinetics of the reaction between the additive and the metallic surface may be also investigated by determining directly the thickness of the reaction layer as a function of time [153]. Thus, for the reaction of a sulphur compound with a steel surface, the reaction product is iron sulphide

$$Fe + (R)_2S \rightarrow FeS. \tag{6.20}$$

As the resulting sulphide layer is subjected to wear, at any particular time its thickness depends both on the formation and the wear rate, i.e.

$$\frac{dC}{C_\infty - C} = (k_1 l + k_2)\, dt, \tag{6.21}$$

where

C is the concentration of iron sulphide on the frictional surface at any given time,

C_∞ — equilibrium concentration of iron sulphide on the friction surface,

k_1 — rate of reaction between the additive and the surface material,

k_2 — rate of wear,

l — effective concentration of the additive on the friction surface.

By integrating, equation (6.21) becomes

$$C = C_\infty (1 - e^{-(k_1 l + k_2)t}). \tag{6.22}$$

Experimental determination of the variation in thickness of the reaction layer with time allows calculation of the equilibrium concentration C_∞ and the sum total $k_1 l + k_2$ (Figure 6.5 *a*).

A simple and rapid method for measuring the thickness of the reaction layer uses radioactive tracers. As the additive used is synthetized with S^{35}, the

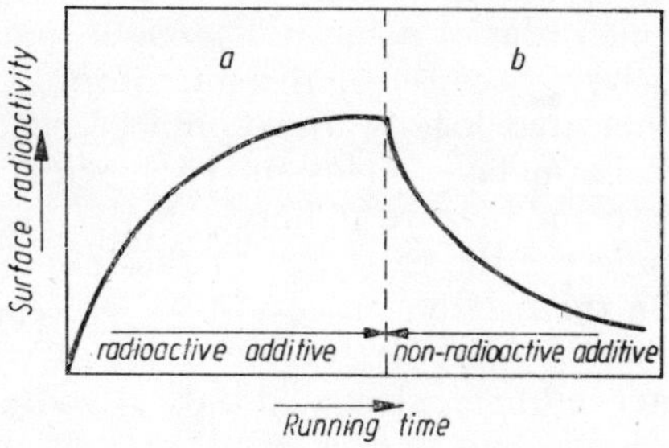

Figure 6.5 Dynamics of reaction layer.

concentration of iron sulphide on the friction surface, assuming that this compound is the only reaction product, is proportional to the surface activity. The constant of proportionality may be determined by calibration.

Another advantage of the radio-tracer technique is the possibility to determine directly the rate of wear of the layer, i.e. the constant k_2. Replacement of the radioactive additive with the same additive, but non-radioactive additive results in a decrease in the surface activity (Figure 6.5 *b*). k_2 is determined from the slope of the curve when the additive is replaced. It should be noted that the use of a classical analytical method allows no separate determination of k_2, as the concentration of the reaction layer does not vary after equilibrium

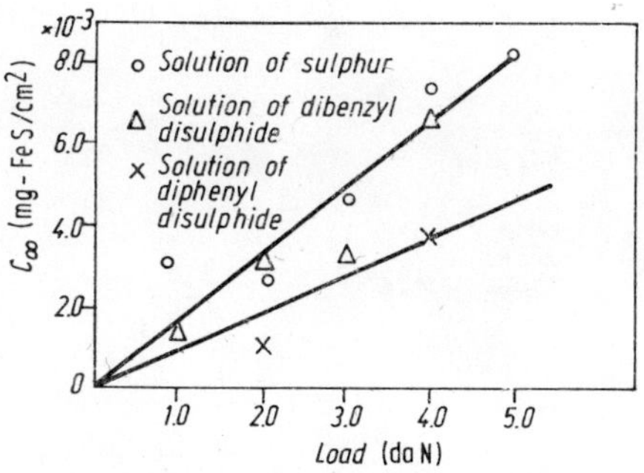

Figure 6.6 Variation of iron sulphide equilibrium concentration on the surface with load [153].

is reached. The radioactive tracer method allows a distinction to be made between the layer formed before and after the additive is replaced.

The values of the constants k_1, l, k_2 and C_∞ as a function of load, enable the effectiveness of the additives in the friction process to be predicted. Thus, the equilibrium concentration of iron sulphide depends on the load and additive type (Figure 6.6).

6.3 Effect of active substances on the friction process

The simultaneous presence of the lubricant molecules, the (anti-wear or anti-friction) additive and surface displacement provides a hydrodynamic effect. The lubricant film thickness, even for heavy loads on non-conformal surfaces is at least one order of magnitude greater than that of a monomolecular adsorbed layer. However, because of the microtopography of real surfaces, the lubricant film is penetrated locally and contact occurs through the adsorbed boundary layer or reaction layer. This mixed lubrication condition is most important in surface wear and seizure (Chapter 6.4).

Considering the existence of the lubricant film it follows that the effect of additives on the friction process in the presence of thick films is negligible. Thus, the coefficient of friction for a friction pair with concentrated load is affected by the presence of the oleic acid only at sliding speeds less than 12 cm/s regardless of the load (Figure 6.7). However, although this effect on the coefficient of

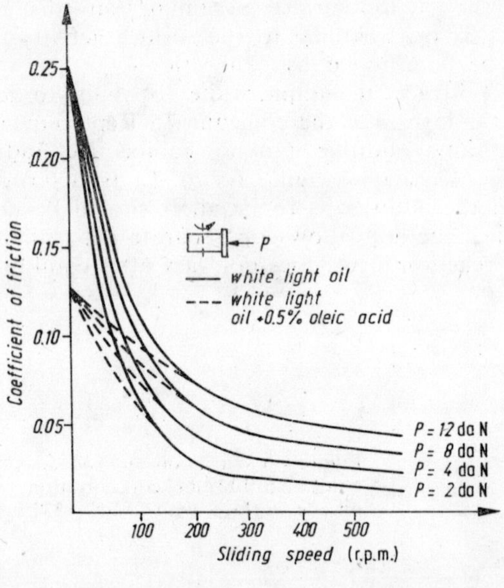

Figure 6.7 Effect of oleic acid on the coefficient of friction under boundary and mixed lubrication conditions. Bush diameter: ∅ = 22.5 mm.

friction occurs only at speeds below 12 cm/s the surfaces are partially separated by the boundary layer at higher speeds. The effect of the adsorbed layers on the coefficient of friction may, however, be established only if the number of contacts becomes sufficiently large.

To reduce the hydrodynamic effect as much as possible, low sliding speeds are generally used for examining the boundary layer. Such operating conditions maintain a low temperature on the friction surface. Temperature, however, is an important quantity characterizing the boundary layer. The importance of the melting point for the effectiveness of homogeneous layers and the importance of the critical desorption point for liquid lubricants are well established. For non-additive lubricants, Matveevski determined the critical desorption point by using a four-ball machine and by heating the lubricant progressively [119]. The low sliding speed (1 r.p.m.) completely eliminates any hydrodynamic effect while slightly affecting surface temperature. He found that between 150°C and 200°C, depending on the lubricant used, the coefficient of friction increased sharply. This was interpreted as a desorption of the lubricant molecules on the surfaces of the friction pair. Under conditions of high speed this surface temperature would cause seizure, according to the theory of critical temperature.

Reduction of the hydrodynamic effect may also be achieved by a reduction in the viscosity of the lubricant. Thus a mixture of hydrocarbons of low molecular weight may be used as an additive solvent instead of mineral oils. The advantage of these hydrocarbons as fluid carriers is that they resemble lubricating oils in their behaviour with additives as far as solubility is concerned and also in their behaviour with metallic surfaces [175, 91]. Because of the increased effect of the boundary layer in this case it is expected that, when hydrocarbons of low molecular weight are used, the effect of the active substances on the frictional process is much more important than with ordinary lubricants. Due to the same reasons the effect of impurities in the liquid used as a lubricant will also increase.

The effect of lubricant purity on the coefficient of friction and hence on the frictional process was examined using an apparatus with three linear contacts [90]. The liquid used as a lubricant was a mixture of normal hydrocarbons with a distillation range from 80°C to 120°C. The friction pairs were manufactured from a low alloy steel containing molybdenum and chromium.

Despite careful measurements being made under constant working conditions, the coefficients of friction were initially found to vary between 0.18 and 0.95 but generally around 0.2. The presence of 0.2% oleic acid stabilized the measured values in addition to reducing the coefficient of friction (Figure 6.8). As the liquid used as a lubricant was a commercial product of limited purity it was assumed that the scatter in the results could be attributed to trace content of active substances resulting from processing or handling. To check this assumption several tests were performed in which the only variable was the method of purifying the lubricating fluid. The results are shown in Figure 6.9. In each test the coefficient of friction was determined for various loads starting with a minimum load.

No special emphasis was laid on the purity of the lubricant during the first 14 runs. In addition to the values in the range from 0.18 to 0.25, to which most runs produced, much greater values of the coefficient of friction such as 0.5 or even 0.95 were also measured. The few high values of the friction coefficient were considered to be characteristic of a pure lubricant. Consequently, the following runs attempted to obtain high values for the coefficient of friction by purifying the lubricant by various methods. Tests 15—33 show an increase in the frequency of high values. Finally, coefficients of friction between 0.8 and 0.9 could be obtained constantly even under low loads.

For further confirmation the lubricant was then deliberately contaminated. Thus, it was found that the addition of 0.001% acetic acid reduced the coefficient of friction from 0.8 to 0.3 and then to 0.24. Smaller amounts of active substance led to instable intermediate values.

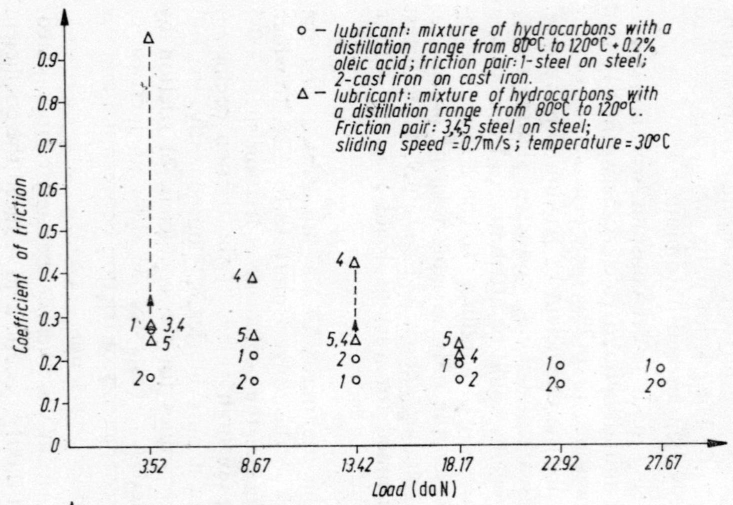

Figure 6.8 Coefficient of friction versus load.

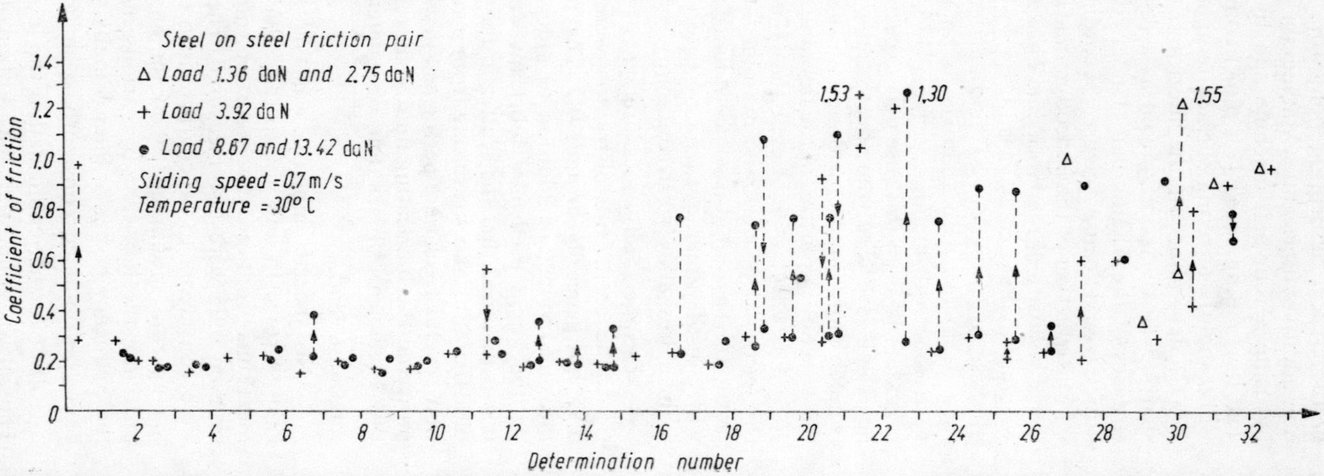

Figure 6.9 Coefficient of friction versus load for several experiments. Lubricant: a mixture of hydrocarbons with a distillation range from 80°C to 120°C.

Low viscosity lubricating fluids may be used to determine the anti-wear properties of additives. The absence of any hydrodynamic effect allows additive properties to be determined for use in frictional processes. Using a linear contact apparatus and a mixture of hydrocarbons with a distillation range from 90°C to 120°C, the anti-wear characteristics of several substances used as additives were determined (Table 6.2). It was found that under moderate loads

Table 6.2

Wear rate for several additives determined in the presence of a lubricant with small molecular weight

No.	Additive	Concentration %	Operating time (h)	Wear rate mg/h
1	without	—	1/6	0.48
2	MoS_2	3	0.5	0.016
3	oleic acid	0.5	1	0.005
4	phosphorus base additive	0.5	1	0.011

Friction pair: tool steel bush on tool steel plate.
Sliding speed: 0.47 m/s
Total load: 4 daN

and speeds oleic acid was the most effective followed by the phosphorus base additive. Use of the fluid as a pure lubricant resulted in a wear rate approximately 100 times higher than for oleic acid.

To provide maximum efficiency a certain relationship should exist between the chain length of the carbon atoms of the lubricant molecules and the number of carbon atoms of the fatty acid used as an additive [14]. This relationship was established on a four-ball apparatus in which the seizure loads for several pure and additive containing hydrocarbons were determined. The lubricants used were pure hydrocarbons with a single type of molecule, namely decane, tetradecane and hexadecane. For these liquids, the seizure load was determined as a function of the number of carbon atoms of the fatty acid (Figure 6.10). A maximum seizure load was always obtained when the number of carbon

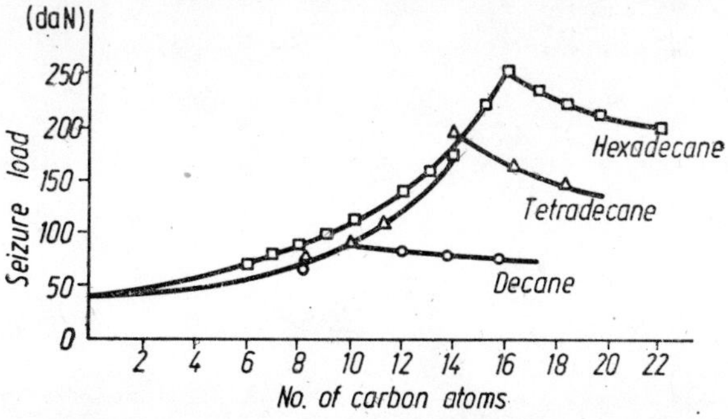

Figure 6.10 Seizure load versus number of carbon atoms in the fatty acid molecule, for several lubricants [14].

atoms in the lubricant and the additive was the same. This behaviour may be explained by the interaction between the two types of molecules (lubricant and additive) which contribute to the formation of the boundary layer. The boundary layer has a maximum strength only if the two types of molecules are similar.

7 Additives and lubricants

Additives are used mainly to improve the performance of lubricants. A class of such substances produces a physical or chemical effect on the surfaces of the friction pair thus leading to either a reduction in the coefficient of friction and wear rate under conditions of mixed or boundary lubrication or an increase in the seizure load. Such additives fall into two groups:

(a) additives used to reduce the coefficient of friction and wear (under conditions of mixed and boundary lubrication);

(b) extreme pressure additives used especially to increase the seizure load of mineral lubricants.

The lack of a precise classification of these additives is due to the various criteria of classification such as effect, operating range and chemical structure used by several authors and to differences in terminology. A correct classification in this field would be possible only when the mechanism of action of several classes of substances is clearly defined in correlation with the frictional process. This problem is still being investigated.

7.1 Additives used to reduce the coefficient of friction and wear rate

This class of additives includes surface-active substances capable of forming a layer on the surface of the friction pair which has sufficient cohesion to withstand the pressure developed in the area of contact. These substances consist of long chain carbon atoms to which an active group is attached to ensure that the molecule has a certain polarity. The most frequently used additives of this type are the esters of the fatty acids, the long chain aliphatic alcohols and amines. Vegetable and animal oils are a source of such substances for industrial use.

The mechanism of action and the properties of the layers formed by these substances are described in Chapters 4 and 6. It should be noted that although these additives are used to reduce the friction coefficient and are sometimes termed accordingly, they also bring about a reduction in surface wear compared with straight lubricants. A characteristic of these layers determined by their structure is that they have a limited stability when the temperature increases. At relatively low temperatures up to 150°C, the layer is desorbed, losing its effectiveness. There are numerous instances (gears, metal machining) when the instantaneous temperature on the frictional surface reaches high values. Under such conditions, additives of the second class are effective as they ensure surface separation by reaction products.

7.2 Extreme pressure additives. Classification

These additives have been developed to improve the performance, especially the scuffing load, of gear lubricants. These additives depend on their chemical composition and fall into the following classes:

(a) *Sulphur additives.* It was found that sulphur, even dissolved in the lubricant, prevented scuffing of highly stressed gears. Sulphur was also combined with several vegetable or mineral oils and its concentration in the lubricant was increased. Among the well-defined sulphur compounds possessing extreme pressure properties are:

— Mono- and disulphides. For example

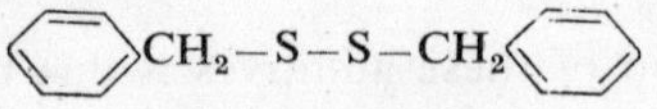

(dibenzyl disulphide)

— Sulphurized terpenes. For example

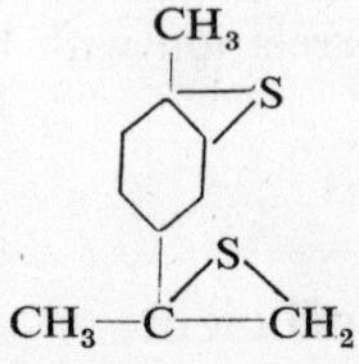

(limonensulphide).

— Sulphurized esters of fatty acids. For example

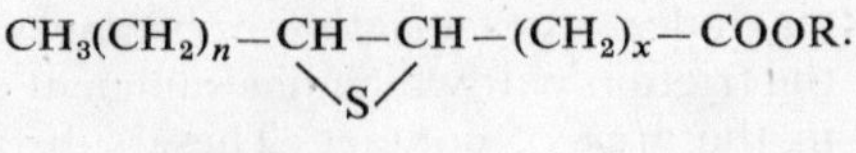

— Xanthates.

(b) *Chlorine additives.* Various chlorine additives are used, the most popular being the hydrocarbon substitution products.
For example, chlorinated hydrocarbon

$$C_nH_{2n-m}Cl_{m+2}.$$

Resistance to hydrolysis is a desirable characteristic of these chlorine additives to prevent highly corrosive HCl being formed by hydrolysis.

(c) *Phosphorus additives.* The most popular phosphorus additives are the esters of phosphoric acid

$$(RO)_3PO, \text{ where } R \text{ is aryl or alkyl.}$$

For example

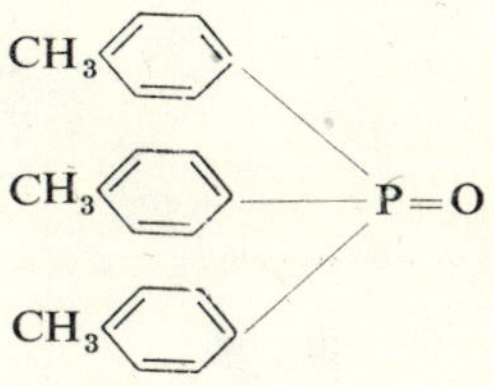

(*p*-cresyl phosphate).

(d) *Additives with two or more active elements in their molecules.* It has been found that the effectiveness of several additives increases and the adverse effects are minimized if two active elements such as sulphur and chlorine, sulphur and phosphorus, sulphur, phosphorus and a metal, are combined.

Within this last extremely large group of additives the class known under the shortened name of metal-dithiophosphate has largely been investigated and used. The most general form of these compounds has the structure

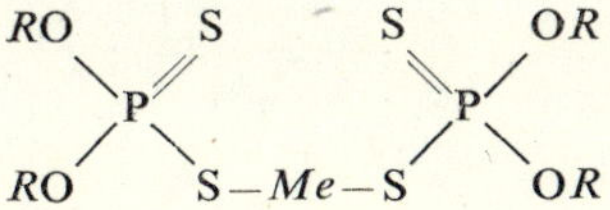

(Me = metal)

where *R*, the same or different, usually are normal or branched alkyl radicals.

Some of these additive materials also have antioxidant and antiwear characteristics in addition to extreme pressure characteristics (e.g. Zn—dithiophosphate). E.P. additives, which usually incorporate active radicals in their molecule, should be used with caution to prevent possible corrosion effects. In several instances, the use of anti-corrosion additives may compensate for the corrosive effects of E.P. additives.

Besides well-defined compounds, mixtures of two or more compounds are generally used. A classical example is the combination of lead naphthenate with a compound containing sulphur in the molecule to provide mild E.P. qualities.

When combinations of additives are used, the antagonistic and synergetic effects should also be taken into account. Thus, when attempting to obtain a lubricant for hypoid gears which needs be effective for both high speed—low torque and low speed—high torque operation conditions, a mixture of two additives, each effective in one of the ranges, does not always provide the desired properties. An additive is required which should function satisfactorily in both ranges simultaneously, such as those in group III (Table 7.1).

Besides efficiency in controlling friction and wear processes other factors require to be taken into account in additive selection. These factors include solubility in the lubricant over a large temperature range, compatibility with other additives present, effectiveness in the presence of other additives, long-term stability and low aggressiveness to metals, rubber and other materials. Formulation of an effective lubricant with well-defined characteristics is laborious and involves extensive laboratory investigation and service testing.

Table 7.1

Extreme pressure characteristics of several organic compounds [157]

	Class of additive	Use in hypoid gears	
		high-speed	high-torque
I	Fatty acids	(−) to −	+
	Esters of fatty acids	−	+
II	Chlorinated organic compounds	+	0 to (+)
	Relatively inactive sulphur in organic combinations without the group COOH	+	(−)
	Active sulphur in organic combinations without the group COOH	+	−
III	Relatively inactive sulphur in fatty acids	+	+
	Relatively inactive sulphur in esters	(−) to (+)	+
	Monoester phosphates	+	+
	Diester phosphates	0	0
	Neutral phosphates	0	0
	Neutral ester thiophosphates	+	+

where

+ very effective,
(+) not very effective.
0 not effective,
(−) slightly harmful,
— very harmful.

7.2.1 The mechanism of action of extreme pressure additives

It is well known that extreme pressure additives react with metallic surfaces during the frictional process. The reaction layer may improve the frictional properties of the metallic surfaces if it is a low shear strength compound or may simply prevent direct contact between the surfaces and the formation of junctions [39]. Sometimes, the resulting layer may aid the retention of lubricant on the surface [111].

Reaction between the metallic surface and the additive may also reduce adhesion. At the points of contact between the surfaces where the temperature is high, the additive prevents the formation of an adhesion bridge by reacting with the metallic surface. As asperities are the initial contact points the process may lead to polishing of the surface (chemical polishing).

Adsorption also plays an important role in the development of this reaction. Prior to reaction, the additive is adsorbed on the surface. The more readily the additive is adsorbed, the higher its concentration on the surface and the greater its reaction rate [164]. The adsorbed layer also ensures separation of the surfaces at low speeds.

An additive may also form a surface polymer layer with good frictional characteristics. The polymer results entirely from a change in the additive. The metallic surface acts only as a catalyst [121]. The surface has a similar passive role when two additives react to form a product with lubricating characteristics [7].

The mechanism of action may be interpreted more correctly if the nature of the layer formed is known. Techniques used for the examination of layers formed by extreme pressure additives include radioactive tracers, elementary X-ray fluorescence analysis, electron microprobe analysis, electron diffraction, chemical analysis and more recently, photoelectron spectroscopy [38, 68, 113, 148, 163, 202].

The radioactive tracer technique uses additives in which one of the active elements is labelled (S^{32}, P^{35}). This technique, although highly sensitive, only establishes the existence of the tracer on the surface without providing any information on the chemical structure of the layer. By successive labelling of various atoms, data on the nature of the layer may be obtained.

Owing to the small amount of products available, elementary analysis should use sensitive techniques to provide information on the chemical structure of the layer. A first indication may be obtained by comparing the relative content of elements in the layer and the additive. The existence of the same relative content of elements shows that the layer is generated by adsorption; changes in the percentage content indicates the occurrence of a reaction.

7.2.2 The reactivity—effectiveness relationship

If the effectiveness of E.P. and anti-wear additives is due to the reaction layer formed on the metallic surfaces in contact, it may be assumed that a direct relationship exists between effectiveness and reactivity. However, additive reactivity should be controlled, that is, the reaction between the metallic surface and the additive should take place only on the friction surface. Excessive reactivity may cause corrosion whilst low reactivity may not permit the formation and preservation of a protective layer on the surfaces of the friction pair as the existence of the layer in the contact area is the result of an equilibrium between the formation and the wear processes.

It is difficult to investigate the reactivity of additives at high temperatures, similar to those reached by metallic surfaces in frictional contact as heating of the whole volume of reactants usually causes secondary reactions which may alter the process under examination. Determination of the reactivity at lower temperatures may be useful only if it is assumed that no other mechanism of reaction occurs at the working temperature. Knowing the constants of the reaction rates at high temperatures a preselection of substances which may act as E.P. or anti-wear additives may be carried out.

Investigation of the reaction between an additive and a metallic surface at high temperature may be carried out without heating all the material, by means of an electrically heated wire [19]. With this technique, the reaction rate is determined by means of the variation of the electrical resistance of the wire. Assuming that the reaction products have a much higher electrical resistivity than that of the wire and that the reaction takes place over the entire wire, the volume of reacted metal may be deduced from the variation of resistance of the wire, before and after the reaction. If the reaction conditions are chosen for the thickness of the reacted layer to be sufficiently small so that the diffusion rate of the metal does not affect the concentration of iron atoms in the reaction area on the one hand, and the concentration of the additive is kept constant on the other hand, the reaction rate will be constant. In this way the reaction constants for several chlorine, sulphur and phosphorus compounds were determined at various temperatures.

The results obtained were correlated with the effectiveness of the respective additives determined as Mean Hertz Load. The values obtained for several additives were compared with those for the pure lubricant (Figure 7.1).

A good correlation between effectiveness and reactivity exists for chlorine additives. No clear relationship exists between the quantities under examination for sulphur and mixed additives, their effectiveness being much greater than that of chlorine additives for the same reactivity. It was found that the reactivity of Zn-dithiophosphate, determined by the same method, is very low, even lower than that of the lubricant. Despite this, its effectiveness as an E.P. or anti-wear additive is comparable to or better than that of sulphur additives. Although

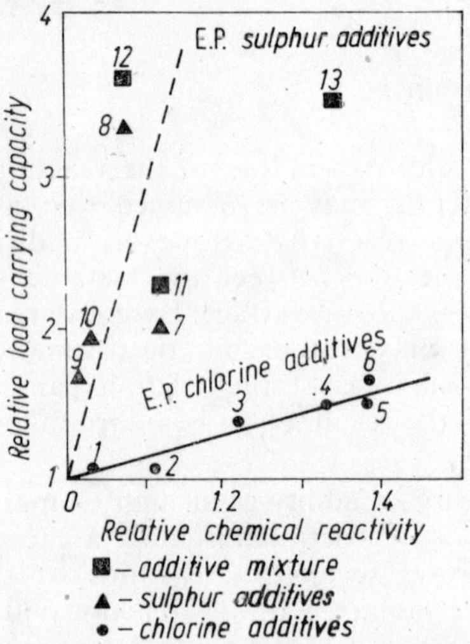

Figure 7.1 Load carrying capacity—reactivity relationship for several sulphur and chlorine additives [154]:

1, benzene chloride; *2*, pentachlorodiphenyl; *3*, methylpentachlorostearate; *4*, chlorinated paraffin; *5*, hexachlorethane; *6*, methyltrichlorostearate; *7*, dibenzyldisulphide; *8*, sulphur; *9*, didodecylsulphide; *10*, diphenyldisulphide; *11*, dibenzyldisulphide + pentachlorodiphenyl; *12*, sulphur + + pentachlorodiphenyl; *13*, sulphur + hexachlorethane.

in certain instances reactivity may be a measure of additive effectiveness the relationship is not generally valid. Reactivity is a necessary property but the nature of the additive and the characteristics of the layer formed also play an important role.

The chemical structure of the additive may affect the formation of the reaction layer if in the molecule, in addition to the reacting atoms (Cl, P, S) groups affecting the adsorption energy of the additives are also present. As the reaction takes place in a heterogeneous system, the kinetics of the reaction is determined, amongst other factors, by the concentration of the additive on the solid surface and not directly by its concentration in the lubricant volume. The concentration of the additive on the friction surface is directly related to the energy of adsorption, a quantity which is considerably affected by the presence of the groups –COOH, –OH, etc. in the molecule. According to this assumption, two molecules containing, for example, the same number of chlorine atoms

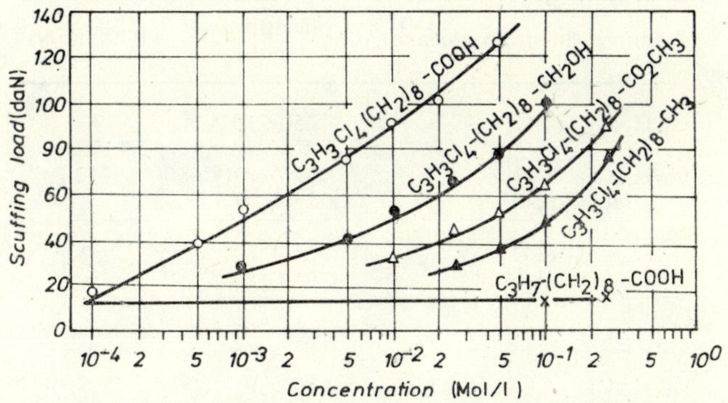

Figure 7.2 Effect of polar groups on load-carrying capacity of several chlorinated compounds under E.P. conditions [164].

but which differ by the presence of a carboxyl group, will have different effects. This assumption was checked on a four-ball machine for several chlorine additives. Additive effectiveness was measured by the 5 seconds seizure load [164]. Figure 7.2 shows that the effectiveness of additives increases if several groups which increase adsorption are attached to the molecule. In all cases the molecules have the same number of chlorine atoms occupying the same position.

The process of adsorption is particularly important for combinations of additives. It is known that for several components, the adsorbed layer contains the available components, their concentration depending upon the energy of adsorption of each component. Thus the presence of a fatty acid in the solution will reduce the efficiency of a compound such as 1, 1, 1, 3-tetrachlordodecane more significantly than that of hydrochloro-lauric acid (Table 7.2). The dislocation capacity of polar substances which are sometimes used as anti-corrosion additives should be taken into account when formulating E.P. lubricants [85].

The effect of combinations of additives on the frictional process may differ from the sum of the effects of each additive. If the effectiveness of the combination is less than that of each component, an antagonistic interaction occurs. If the effectiveness is higher, a synergetic interaction occurs. The antagonistic effect of sulphur additives with various polar substances was determined by the direct examination of the layer formed on gear flanks. It was found that the reaction between a lubricant with a sulphur content of 0.65% and the surface of the tooth flank was prevented by the presence of lauryl phosphate in a concentration of only 1% [20]. This explains why E.P. additives are more effective in highly purified mineral lubricants compared with the same lubricants, subjected only to an ordinary refining process [149].

Table 7.2

Effect of several surface-active substances on the effectiveness of chlorine additives]164[

Additive (in *n*-hexadecane)	Concentration, mol/l	P(*) (daN)
$C_{11}H_{19}Cl_4-CH_3$	$250 \cdot 10^{-3}$	64
$C_{11}H_{19}Cl_4-CH_3$ + lauric acid	$250 \cdot 10^{-3}$ $250 \cdot 10^{-3}$	12
$C_{11}H_{19}Cl_4-CH_3$ + lauric alcohol	$250 \cdot 10^{-3}$ $250 \cdot 10^{-3}$	48
$C_{11}H_{19}Cl_4-CO_2CH_3$	$160 \cdot 10^{-3}$	64
$C_{11}H_{19}Cl_4-CO_2CH_3$ + lauric acid	$160 \cdot 10^{-3}$ $160 \cdot 10^{-3}$	22
$C_{11}H_{19}Cl_4-CH_2OH$	$50 \cdot 10^{-3}$	64
$C_{11}H_{19}Cl_4-CH_2OH$ + lauric acid	$50 \cdot 10^{-3}$ $50 \cdot 10^{-3}$	26
$C_{11}H_{19}Cl_4-COOH$	$5 \cdot 10^{-3}$	64
$C_{11}H_{19}Cl_4-COOH$ + lauric acid	$5 \cdot 10^{-3}$ $5 \cdot 10^{-3}$	40
$C_{11}H_{19}Cl_4-COOH$ + lauric acid	$5 \cdot 10^{-3}$ $50 \cdot 10^{-3}$	24
$C_{11}H_{19}Cl_4-COOH$ + lauric alcohol	$5 \cdot 10^{-3}$ $5 \cdot 10^{-3}$	52

(*) Increase in the seizure load with respect to pure hexadecane determined on a four-ball machine.

7.2.3 **Sulphur additives**

The simplest method of investigating the layer formed by sulphur additives is to use sulphur in the form of a solution in the lubricant. Using a SAE test machine, Godfrey investigated the reaction layer and the composition of the wear debris formed when using a saturated solution of sulphur in medicine oil [68]. The main component detected by electron diffraction, both on the surface and in the wear debris was not iron sulphide but Fe_3O_4 or a mixed oxide. Results of spectrographic analysis also showed an increase in the content of silicon present in the form of SiO_2 in the surface layer. The process of concentration of a compound on the contacting surface has been established also in other instances and is attributed to selective corrosion at high temperatures [65, 174].

In addition to iron oxide, iron sulphide was also found to be present by chemical analysis. The maximum sulphur concentration was found to be 10%, which corresponds to a 25% concentration of FeS. The effectiveness of the layer which forms is, however, due to the presence of both FeS and iron oxide. The surfaces of friction pairs from which FeS was removed by treatment with a dilute acid or surfaces with a preformed layer of FeS or Fe_2O_3 scuff under much lighter loads than those having a complete layer.

With organic sulphur compounds, the composition of the layer depends both on the load and the organic radical of the molecule, as reported by Allum from four-ball machine tests. He examined the frictional surface by electron probe micro-analysis [6]. The mean diameter of the wear scar and the content of sulphur and oxygen in the centre of the scar determined by microanalysis (E.P.M.A.) are given in Table 7.3. In all cases the concentration of additives was

Table 7.3

Mean diameter of wear scar and composition of surface layer as a function of additive structure. Load P = 160 daN, speed = 1.500 r.p.m., running time = 60 min (after [6])

Additive (disulphide)	Mean diameter of wear scar (mm)	Sulphur content in the track Preponderant value/max. value (%)		Oxygen content in the track Preponderant value (%)
Di-*n*-dodecyl	2.90	1	6	—
Diphenyl	2.71	11	14	15
Di-*n*-butyl	2.25	8.5	16	19
Di-tert-butyl	1.71	8	18	8
Diallyl	1.15	—	30	7

18.52 mmol per 100 g of paraffin oil. Table 7.3 shows a general tendency for the content of sulphur and subsequently, of FeS, to increase as the additive effectiveness increases. Increase of sulphur content is accompanied by a decrease in oxygen content. This increase in sulphur content in the scar track with increase in additive effectiveness indicates that FeS plays a major role in the effectiveness of reaction layer.

The main advantage of microanalysis for the examination of surfaces consists in the possibility of determining both the point and mean concentration and the surface distribution of the elements under investigation.

Figure 7.3 shows an area of the centre of the wear scar (*a*) and the sulphur distribution over this area (*b*). The element is concentrated in the rougher areas which indicates the existence of an adsorption effect.

Suggesting a relationship between the structure of the sulphur compounds and their effectiveness as E.P. additives, it was concluded that effectiveness is determined by the energy of the S—C bond [4]. The faster this bond breaks, the sooner the reaction of formation of FeS takes place. For monosulphides the S—C binding energy decreases in the series:

diphenyl > di-*n*-butyl > di-*sec*-butyl > di-*tert*-butyl > dibenzyl,

an order which should be applicable to disulphides. This series coincides with that given in Table 7.3, which confirms the assumption made.

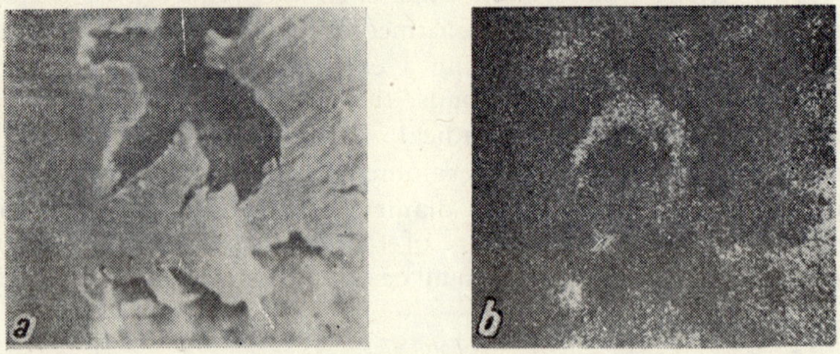

Figure 7.3 Distribution of reaction products for a sulphur additive [59]: *a*, wear scar picture; *b*, picture of sulphur distribution in the wear scar obtained by electronprobe microanalysis.

Under conditions of mild wear (load $P = 15$ daN, speed = 1500 r.p.m., running time = 30 min.) without seizure, a surprising reversal in additive effectiveness occurs, the diphenyl-disulphide now being the most efficient [5]. This behaviour could be accounted for by the tendency of the corresponding mercaptan to form which is determined by the energy of the S—S bond. Considering these results, the following mechanism of action of disulphides is suggested: adsorption of the disulphide on the metallic surface, scission of the S—S bond and formation of a layer of Fe-mercaptan. This layer, which resembles a chemisorbed layer protects the surface effectively if the conditions in the frictional area are not excessively severe. Under heavy loads in extreme pressure conditions the C—S bond breaks and the iron sulphide which forms takes over the role of surface protection.

7.2.4 **Chlorine additives**

As with sulphur additives, it is assumed that the effectiveness of chlorine additives is due to the formation of the corresponding chlorides on the metallic surfaces of the friction pair. Determinations carried out on a layer of iron chloride, formed by the reaction of a metal with gaseous chlorine, showed a reduction in the coefficient of friction especially at temperatures over 150°C (Figure 7.4). It was also found that organic chlorinated compounds are inactive during the frictional process up to temperatures of approximately 200°C [39].

The lack of reaction layer for organic chlorinated compounds under conditions of moderate load and speed was established with a pin on disc machine [96]. No effective measurable layer was reported to form on the surfaces at pressures up to approximately 2,200 kg/cm^2. It is probable that the effect of the chlorinated compounds occurs under heavy loads when, in addition to high temperature, the metallic surface is activated mechanically (tribochemical effect).

It is known that freshly machined surfaces are a source of electrons which are capable of initiating several reactions which would otherwise not occur [74]. Thus, the reaction of CCl_4 with a freshly machined aluminium surface, as occurs

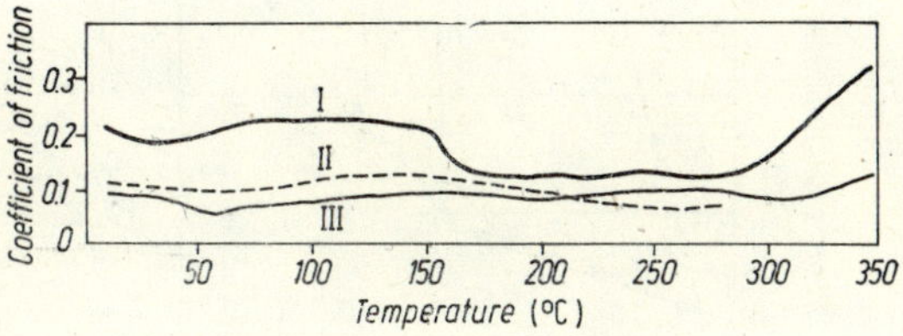

Figure 7.4 Coefficient of friction of the chloride layer on steel surfaces versus temperature [39]: I, chloride layer; II, chloride layer and paraffin oil; III, ferric chloride layer.

in metal cutting, involves

$$Cl_3C^{\delta+} - Cl^{-\delta} + e \rightarrow Cl_3C\cdot + Cl^-,$$

$$Al^{+++} + 3\,Cl^- \rightarrow AlCl_3,$$

$$2\,Cl_3C\cdot \rightarrow Cl_3C - CCl_3.$$

Chloride formation on a steel surface during metal cutting under controlled conditions has been investigated [163]. Various chlorinated compounds, in solution in *n*-hexadecane were used as cutting fluid. The chlorine content in the resulting chips, determined by chemical analysis is shown in Table 7.4.

Table 7.4

Thickness of the reaction layer on the chips resulting from steel drilling, using several chlorine additives in two concentrations (after [163])

Additive	gCl^-/cm^2	
	0.5 mol/l	0.05 mol/l
Ethylene trichloride	$45 \cdot 10^{-8}$	—
Tetrachlorethylene	$54 \cdot 10^{-8}$	$35 \cdot 10^{-8}$
Chloroform	$110 \cdot 10^{-8}$	—
1,1,2,2-tetrachlorethane	$140 \cdot 10^{-8}$	$49 \cdot 10^{-8}$
Carbon tetrachloride	$190 \cdot 10^{-8}$	—
Hexachlorethane	$220 \cdot 10^{-8}$	$80 \cdot 10^{-8}$
Octachlorpropane	$320 \cdot 10^{-8}$	$140 \cdot 10^{-8}$

To compare reactivity and effectiveness of the additives, wear was also determined with a four-ball machine (Figure 7.5). A close relationship is found between the reactivity of the additive, determined by means of the above technique, and its effectiveness, although an appreciable difference exists between the conditions defining the two processes.

The composition of the layer formed by chlorinated additives with a steel surface where such layers are able to form and be maintained, i.e. under static conditions and in metal cutting, was found to be $FeCl_2$ or $FeCl_3$.

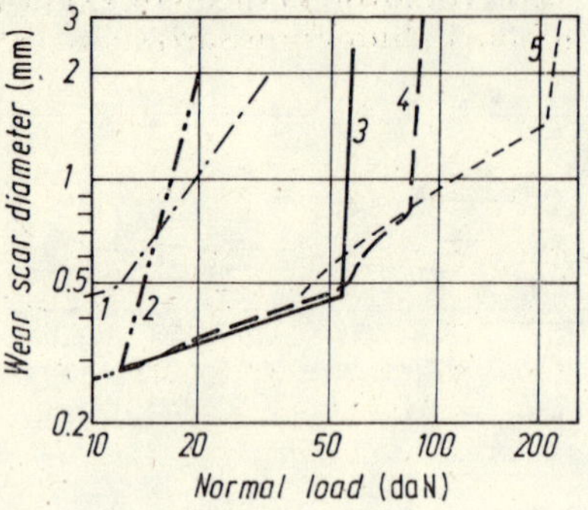

Figure 7.5 Anti-wear and E.P. properties determined with a four-ball machine for several chlorine additives [163]:

1, dry; *2*, *n*-hexadecane; *3*, tetrachlorethylene; *4*, tetrachlorethane; *5*, octachlorpropane; chlorinated compounds in concentration of 0.5 mol/l in *n*-hexadecane.

7.2.5 **Phosphorus additives**

In the very comprehensive class of phosphorus additives the esters of phosphoric acid were especially investigated from the point of view of their mechanism of action.

An early explanation proposed for the mechanism of action of tricresyl phosphate included the formation of the iron-iron phosphide eutectic, providing a surface polishing effect. Subsequent investigations did not confirm this assumption, the iron phosphide being identified only with static corrosion [19].

At present, it is generally assumed that the layer formed by tricresylphosphate on steel surfaces consists of ferric or ferrous phosphate [71, 163, 32]. Recent studies on the reaction of tricresylphosphate with steel surfaces of a pin

Table 7.5

Wear of a steel surface in air and azote in the presence of tricresylphosphate [71]

Atmosphere	Load (daN)	Speed (cm/s)	Pin wear (mm^3)		Surface aspect
			wet	dry	
Air	25	50	0.004	0.005	low roughness
Azotc	25	50	0.007	0.018	strong adhesion
Air	50	25	0.009	0.012	low roughness
Azote	50	25	0.043	0.038	strong adhesion

on disc pair established by electron diffraction the presence of the compounds $FePO_4$ and $FePO_4 \cdot 2\,H_2O$. After the layer has formed, the surfaces acquire a characteristic feature being covered with a white crystalline substance readily observed by dark field microscopy.

The presence of oxygen and water plays an important rôle in the development of the reaction between tricresylphosphate and a steel surface (Table 7.5).

The influence of the water and oxygen indicates that the formation of phosphate is preceded by hydrolysis of the ester and release of phosphoric acid which finally reacts with the oxide layer on the surface.

Table 7.6

Anti-wear properties of phosphoric acid compounds determined with a four-ball machine (after [30])

Additive	Concentration % weight	Mean diameter of wear scar (mm)		
		P = 1 daN	P = 10 daN	P = 40 daN
Superrefined mineral oil $\nu = 75$ cSt at 25°C				
	non-additivated	0.25	0.47	0.59
Tricresylphosphate	0.01	0.17	0.25	—
	0.5	0.17	0.24	0.97
	1	0.17	0.24	0.47
Lauryl acid phosphate	0.05	0.17	0.24	0.46
Di-3-methylbutyladipat				
	non-additivated	0.39	0.71	0.91
Tricresylphosphate	1	0.38	0.71	0.97
	3	0.40	0.64	0.97
	5	0.23	0.25	0.72
Lauryl acid phosphate	0.01	0.21	0.41	0.84
	0.05	0.19	0.28	0.43
Phosphoric acid	0.001	0.41	0.69	0.90
	0.01	0.16	0.37	0.50

The importance of the source of the tricresylphosphate used (ordinary commercial or highly purified) as well as the atmospheric environment have led to investigations concerning the role of impurities on the effectiveness of this additive [30]. By using thin layer chromatography for analysis and preparative chromatography on iron powder to purify the product it has been shown that the presence of several polar impurities have significant effects on the frictional process when using commercial tricresylphosphate. Table 7.6 lists the anti-wear properties of several derivatives of phosphoric acid in various basic lubricants.

From Table 7.6 the acid ester, an impurity, in a concentration of only 0.05% has almost the same effect as tricresylphosphate in a concentration of 1%. If the basic lubricant is properly selected phosphoric acid appears to be the most effective. In conclusion, it was suggested that the tricresylphosphate represents only the source of several more effective compounds such as acid esters or even phosphoric acid which would form by hydrolysis, oxidation or thermal decomposition. Esterification of phosphoric acid is necessary only to ensure solubility of the product in mineral oils.

Another widely used class of additives are the dithiophosphates. These additives used in the form of neutral salts, usually as Zn-dithiophosphates are considered to be polyfunctional. They are used as oxidation inhibitors, anti-wear and anti-corrosion additives.

The effect of chemical structure on the effectiveness of dithiophosphates in the frictional process has been investigated in some detail [7, 59, 151]. The effect of the nature of the metal and of the organic part on the wear rate, surface topography and the composition of the reaction layer have been studied for a large range of additives. The anti-wear or extreme pressure properties for various salts of metal-di(4-methylpenthyl-2) dithiophosphate determined with a four-ball

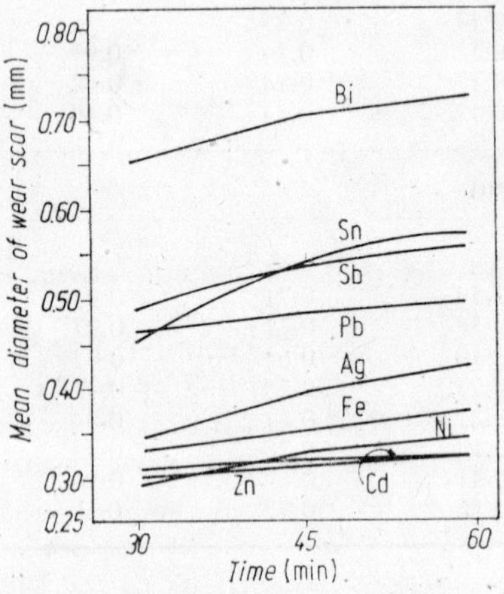

Figure 7.6 Effect of the metal in M-*di*-(4-methylpenthyl-2) dithiophosphate additive under mild wear conditions determined with a four-ball machine [7], load $P = 15$ daN, temperature $\theta = 50°C$.

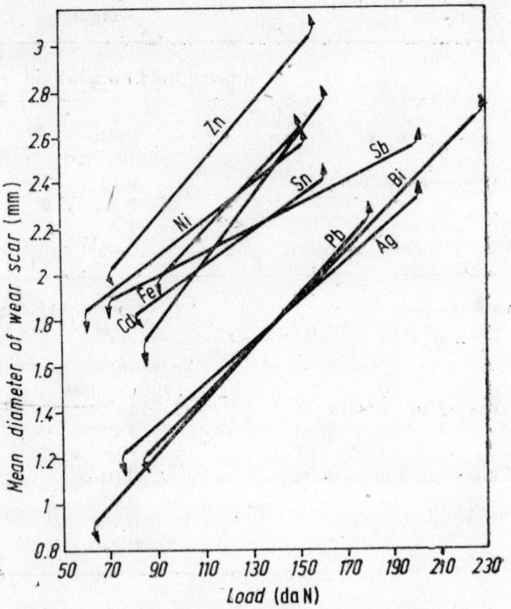

Figure 7.7 Effect of the metal in M-*di*-(4-methylpenthyl-2) dithiophosphate under extreme pressure conditions determined with a four-ball machine [7].

machine under mild and severe wear conditions after seizure are shown in Figures 7.6 and 7.7. The major effect of the metal is noted in both operating regimes, although in a different order. Thus, under extreme pressure conditions additive effectiveness decreases in the following order

$$\text{Ag, Bi, Pb} > \text{Sb} \geqslant \text{Sn} > \text{Cd} > \text{Fe, Ni} > \text{Zn}.$$

Under light loads it decreases in the order

$$\text{Zn, Cd, Ni} > \text{Fe} > \text{Ag} > \text{Pb} > \text{Sb, Sn} > \text{Bi}.$$

As with disulphides, it appears that dithiophosphates also exhibit a reversal of properties in the two operating regimes.

The effect of the metal on the anti-wear properties was also established directly by determining wear in a cam-tappet assembly. It was reported that Zn and Cd salts are more effective than Ni salts [28].

In an attempt to interpret the effectiveness of various dithiophosphate compounds, several authors have attempted a correlation with the thermal stability of the product. This can be readily determined by means of the temperature at which the release of H_2S and olefines occurs. This however is only a small contribution to the elucidation of the mechanism of action of these additives.

Analysis of the wear scar by electron microprobe analysis reveals that under light loads ($P < 40$ daN with a four-ball machine) the most effective additives form the highest metal and phosphorus concentrations. It is probable that the effectiveness of these additives is due to the formation of a layer of phosphate or thiophosphate in which the organic radicals are still present. The Ni, Zn and Cd salts form layers in which most of the sulphur existing initially in the additive is lost.

Under extreme pressure conditions the most effective additives are Bi and Pb salts which produce the lowest metal, sulphur and phosphorus concentrations in the wear scar. Though no explanation for this has yet been found, two distinct mechanisms of action probably exist.

As with disulphides, the maximum concentration of the elements which make up the layer occurs in the roughest areas. Comparison between the microscopical image of the surfaces and the distribution of the main elements (P, S, Cd) shows that the latter follows faithfully the defects in surface topography. This conclusion may render the effectiveness of the reaction layer questionable, as to the extent it can act if it lies in surface depressions, that is out of the contact area.

A relatively simple technique for the detection of reaction layers is to measure the contact electrical resistance. This technique, mainly used for examining the lubricant film, may also under certain conditions be used to detect the presence of reaction layers [96]. A pin on disc pair and associated apparatus were used to obtain the simultaneous measurement of the most important parameters of the frictional process such as normal force, frictional force, oil and surface temperature and contact electrical resistance [16, 17]. To eliminate the effect of the fluid lubricant film, the sliding speed was decreased when the frictional force and the electrical contact resistance of the reaction layer were determined ($V = 0.1$ cm/s and $V = 0$ cm/s, respectively). Values measured in this way depend on the properties of the surfaces, i.e. the properties of the layers. An additional condition required in order to identify the presence of a reaction layer by such measurements is that the layer and the metallic surface must have different resistivities and coefficients of friction. As adsorbed layers are very thin they cannot be identified by measuring the electrical contact resistance, but they can be readily detected by their effect on the coefficient of friction.

To obtain a reaction layer which should be measurable by the above technique, the following conditions should be obeyed:

(a) the formation rate should be higher than the wear rate;

(b) the reaction products should be insoluble in the lubricant; they should possess adequate mechanical strength and adhesion to the frictional surface;

(c) the reaction time should be sufficient to ensure the formation of a continuous reaction layer;

(d) the operating conditions should be chosen so that severe wear does not occur.

The last item is particularly important as under high loads, after seizure has occurred and the surfaces have become rough, it is not possible to form a measurable continuous layer.

Table 7.7

Coefficient of friction (μ) and electrical contact resistance (R) for pure and additivated mineral oil. Friction pair: pin on disc [96]

Measured quantity \ Additive	Pure oil	1% oleic acid	1.5% dibenzyl disulphide	1.5% dithio-phosphate	1.5% chlorinated paraffin
μ (unwashed)	0.14	0.11	0.2	0.1	0.2
$\bar{R}$ (unwashed) [Ω]	5.10^4	20	1	10^5	200
μ (washed)	0.5	0.55	0.55	0.11	0.5
$\bar{R}$ (washed) [Ω]	5.10^4	3	2	10^5	50
μ (before running)	0.22	0.11	0.2	0.11	0.2
$\bar{R}$ (before running) [Ω]	1	1	1.4	1	1

where μ was measured at $V = 0.1$ cm/s, R was measured at $V = 0$.

Table 7.7 gives the contact resistance and the coefficient of friction both in the presence of the lubricant and after the pair has been washed with petroleum ether to remove the adsorbed layer. The values listed were determined on the reaction layers obtained under a load of 140 daN and a reaction time of 2 hours. The contact resistance given in Table 7.7 is the average of at least 5 measurements owing to the wide spread of this quantity. For comparison, the table also gives the coefficients of friction and contact resistances of the initial surfaces in lubricated conditions ($t = 0$).

Under the above conditions only Zn-dithiophosphate and pure vaseline oil formed reaction layers. The formation of a layer with pure mineral oil was attributed to surface oxidation which was confirmed by microscopical examination. Although this layer has a high coefficient of friction under dry conditions, in the presence of a lubricant the coefficient of friction approaches that of oleic acid. This behaviour may be attributed to the high adsorption capacity characterizing the oxides. The other additives, namely oleic acid, dibenzyl disulphide and chlorinated paraffin did not form measurable reaction layers under the conditions used. Oleic acid forms an adsorbed layer of known low coefficient of friction. The physical nature of the bond between this layer and the surfaces of the pair follows from the fact that it is readily removed by washing.

Figures 7.8 and 7.9 are scanning electron micrographs of a pin showing the formation of the oxide layer or the onset of seizure. It is evident from Figure 7.9 that a continuous reaction layer is not possible on such surfaces.

7.2.6 Reaction layers without the contribution of surfaces

In addition to the reaction between additive and the friction surface there is another possibility to form protective layers on the friction surface. The additive

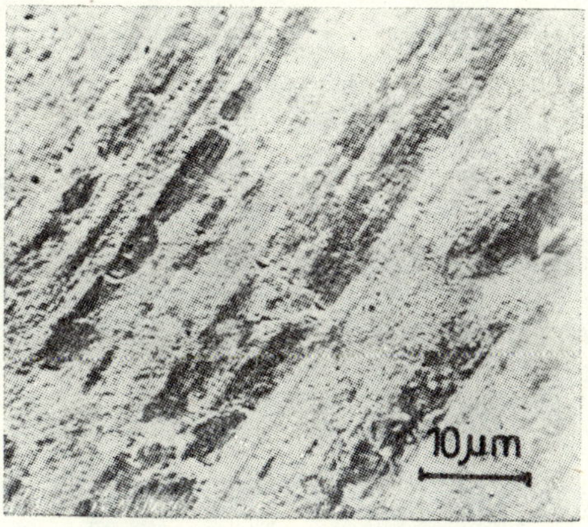

Figure 7.8 Formation of a reaction layer a steel surface. Picture obtained in a scanning electron microscope.

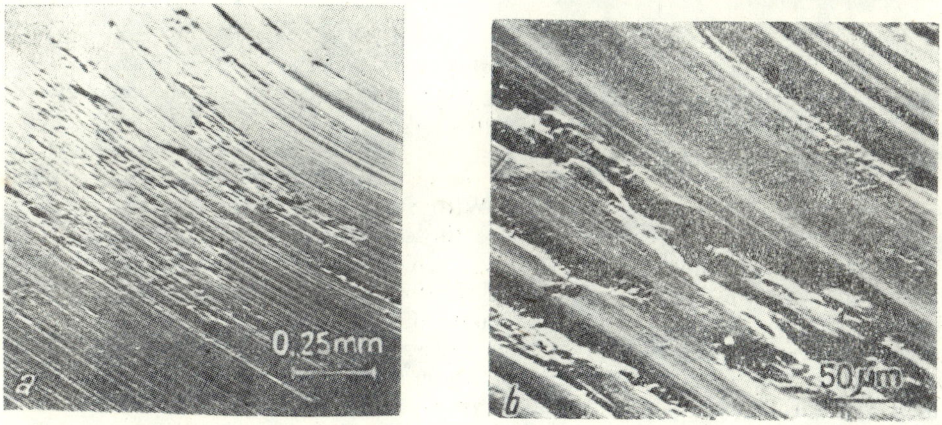

Figure 7.9 Surface of a steel pin in the first stage of seizure. Pictures obtained in a scanning electron microscope.

or additives present in the lubricant, in contact with the frictional surface at high temperatures undergo polymerization or reaction with one another leading to the formation of a solid compound on the surfaces [86].

It has been found that Zn-dithiophosphates may suffer similar reaction and form polymers on the friction surface. The reaction, which takes place only at a certain temperature, has two stages [121]:

1) 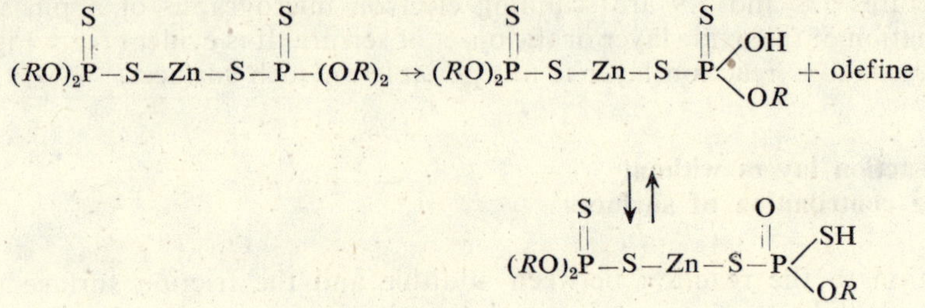

2) 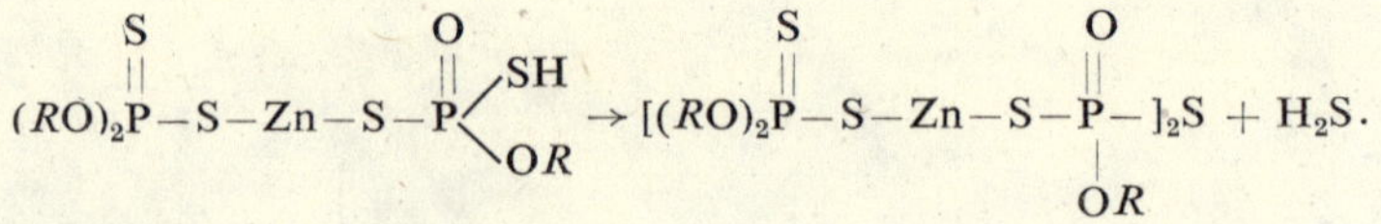

The reaction continues with the formation of a macromolecule. This process begins at 176°C and its rate increases rapidly with temperature. Thus, the *in situ* polymerization of Zn dithiophosphate, which occurs by a temperature rise, may be another mechanism of action of this additive. The polymer layer formed affords protection of the metallic surfaces against corrosion.

7.3 **The process of scuffing**

Scuffing, a form of wear, is the rapid deterioration of the surfaces of the friction pair by adhesion. Surface deterioration by scuffing depends on the mechanical characteristics of the surfaces and the intensity and duration of the load. Thus the scuffed surface may have various features. If the surface has a low hardness, adhesion causes separation within the material with probable extension over adjacent areas (Figures 5.8 and 7.9). With surfaces of high hardness scuffing results in characteristic features such as those shown in Figure 7.10.

It is generally agreed that with a given lubricant scuffing occurs because the lubricant film fails locally due to a rise in surface temperature. The temperature rise may occur due to inadequate design, incorrect construction and mounting or operating factors such as excessive speed, load and temperature.

If the causes generating scuffing are not rapidly detected the particular intensity of the process may lead to rapid surface deterioration. Dimensional changes due to scuffing may render further normal operation impossible. For mechanisms operating with small clearance (such as the piston-cylinder assembly) the onset of scuffing may cause complete seizure.

To increase the scuffing limit of surfaces, extreme pressure additives may be used in addition to other methods. Currently used lubricants for heavily loaded gears contain such additives.

Due to progress in gear processing and the development of materials and lubricants, deterioration by scuffing is generally no longer a problem except during the running-in of gears operating under severe conditions. However, surface protection against scuffing is continuously arising nowadays when attempts are made due to the development of gears and engines of advanced design and high ratings.

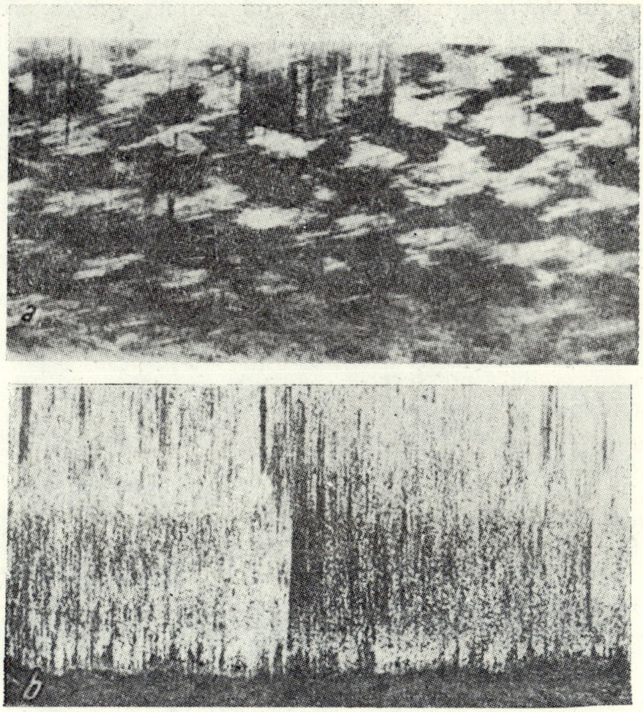

Figure 7.10 Gear flank in different wear stages [25]:

a, mild wear by scratching; *b*, severe wear by scuffing, extended over the whole tooth flank.

7.3.1 Scuffing criteria. The flash temperature concept

As scuffing is initiated by increasing load and speed, some authors have proposed simple empirical scuffing criteria (Almen, Hopler, etc.). These relationships are, however, valid only under the conditions for which they were derived. A critical review of scuffing criteria has been carried out by Pavelescu [137].

Block (1937) made the first attempt to approach the scuffing process from a more general viewpoint; he proposed the flash temperature concept based on the constancy of the flash surface temperature upon scuffing. For a pair of given materials, lubricated with an additive free oil, scuffing occurs if the critical temperature is achieved by increasing the speed or the load. This postulation together with a method for the calculation of the flash surface temperature proposed by the same author permits the determination of the scuffing (seizure) limit based on a single experimental determination [33, 169].

The method of calculating the flash temperature is such that the proposed criterion may be applied to any friction pair. Thus the method may be applied to gearings.

To calculate the flash temperature for gears, relation (3.11) is used, in which the expressions for the mean heat flow q_m and contact width w are introduced. In the case of cylindrical gears these quantities are given by the relations

$$q_m = p_m \mu (V_1 - V_2), \tag{7.1}$$

$$p_m = \frac{W}{w}, \tag{7.2}$$

$$w = \sqrt{\frac{32}{\pi}} \sqrt{\frac{Wr'}{E'}}, \tag{7.3}$$

where

p_m is mean contact pressure,
W — load per unit length,
w — width of the contact band,
μ — coefficient of friction,
r' — reduced radius,
E' — reduced modulus.

The value of 1.11 which is taken for the constant A in relation (3.11) corresponds to an elliptical distribution of heat flow. Substituting relations (7.1), (7.2) and (7.3) into (3.11) and assuming that both gears are made of the same material, the following relation is obtained

$$\theta_{max} = 0.62 \mu W^{3/4} [\sqrt{V_1} - \sqrt{V_2}] \sqrt[4]{\frac{E}{b^2 r}}, \tag{7.4}$$

where

$$b = \lambda \rho c.$$

This temperature rise is added to the mean surface temperature θ_M. The surface temperature is then

$$\theta_{Tmax} = \theta_{max} + \theta_M. \tag{7.5}$$

As in gears the sliding speed, the load and the radius of curvature are variables, it is necessary to calculate the maximum value for expression (7.4). Considering the situation existing in real gears, it is considered that the expression

$$W^{3/4} \frac{\sqrt{V_1} - \sqrt{V_2}}{\sqrt[4]{r}}$$

reaches its maximum value at the initial gear contact. Hence, the temperature will be calculated for the values of V_1, V_2, W and r corresponding to this situation.

To calculate the surface temperature from relation (7.4) the value of the coefficient of friction is required. This quantity is difficult to determine accurately for gears. Use of a mean, approximate value may lead to erroneous results. By modelling the operation of gears on a roller machine, Kelley found that the dependence of the coefficient of friction on a conveniently chosen parameter in a logarithmic representation is a line (Figure 7.11) [27]. However, not all lubricants yield such a straight line with respect to the parameter $\log(\eta_0 V_a/W)V_T^2$. However, changing the exponents of several quantities such as W and η_0 a linear dependence may be obtained.

The tooth mean temperature θ_M may be derived from the gear heat balance, that is from its coefficient and temperature losses considering, for instance, that the gear temperature is equal to the casing temperature. For a normal toothed gear, the heat released Q (kcal/h) may be expressed by the relation [116]

$$Q \leqslant 6\mu \left(\frac{1}{Z_1} \pm \frac{1}{Z_2}\right) C_{\text{dyn}}\, N \cdot 632, \tag{7.6}$$

where

Z_1, Z_2 is number of teeth,
C_{dyn} — dynamic loading coefficient,
N — transmitted power (HP),
"+" or "−" for external, internal gear, respectively.

For determining the load W, the dynamic and deformation effects as wel as all the other effects leading to an increase in the real load as compared to the rated load should be taken into account.

If the critical temperatures for a given lubricant and material are known, the critical load may be deduced from relation (7.4) in terms of tooth geometry and rotational speed.

The validity of the flash temperature concept in the form proposed by Blok has been confirmed experimentally with gear machines and four-ball machines [107, 89] and more recently with a disc machine [109]. The last work showed that the flash scuffing temperature does not depend on the sliding speed or viscosity for a large range of values for these quantities (Figure 7.12). It was also found that the ratio of the calculated thickness of the lubricant film to the asperity height did not determine the onset of scuffing. Cases existed when the surface scuffed even if the thickness of the lubricant film was greater than the highest asperities. Cases also existed when, although the thickness of the lubricant film was smaller than the maximum height of the undeformed asperities, the surface did not scuff. Figure 7.13 plots the variation of the maximum flash temperature at scuffing with the thickness of the lubricant film calculated from the relations provided by elasto-hydrodynamic theory. For film thicknesses from 0.25 μm to 1.3 μm surface scuffing occurred at approximately the same temperature (180°C).

Although these results appear to confirm the basic assumption of a flash temperature concept, the existence of a lubricant film with concentrated load pairs confirmed theoretically and experimentally, contradicts a basic assumption made when relation (3.11) was derived, namely direct contact between the surfaces. Actually with a pair under concentrated load, at a certain speed, a load carrying lubricant film exists, regardless of the load applied. The existence of this film in

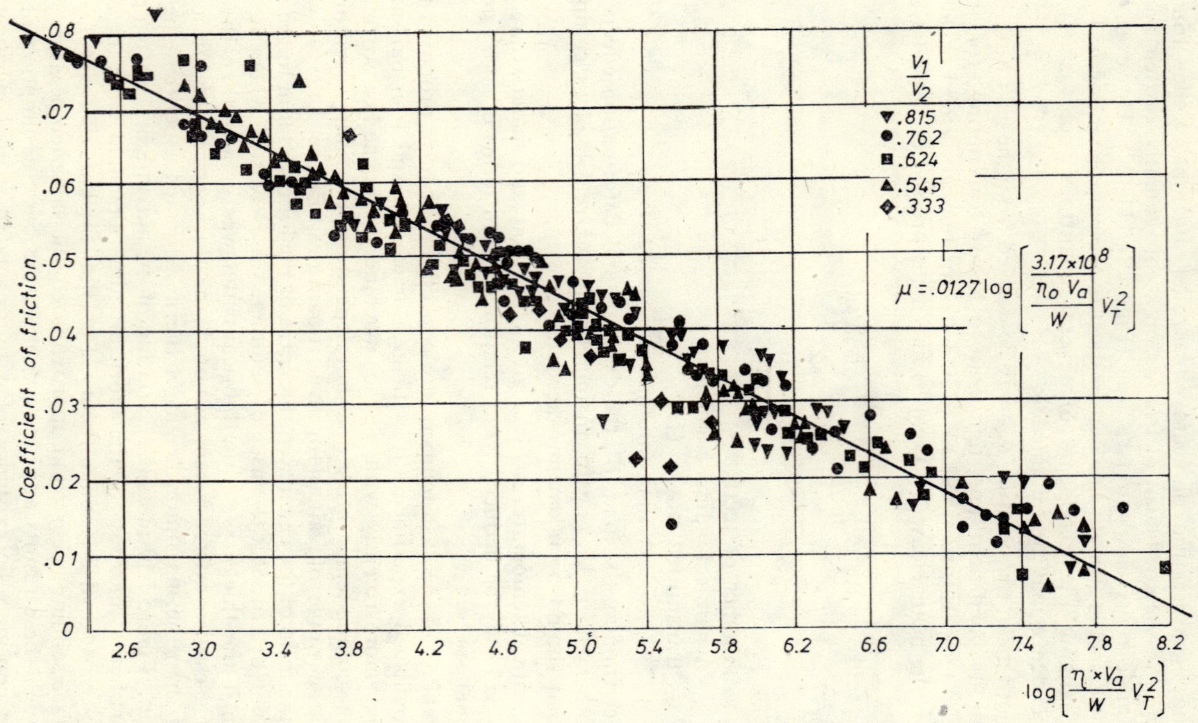

Figure 7.11 Coefficient of friction in gears versus operating parameters [27]: $V_a = V_1 - V_2$; $V_T = V_1 + V_2$. Units of measure: η (cP); V(in/s); W(lb/in).

addition to the oxide film affects the distribution of temperatures in the contact area. The real case is further complicated by the fact that under that normal loads the elasto-hydrodynamic film is continuously penetrated by asperities, thus heat is produced both by viscous friction and by frictional contact between the

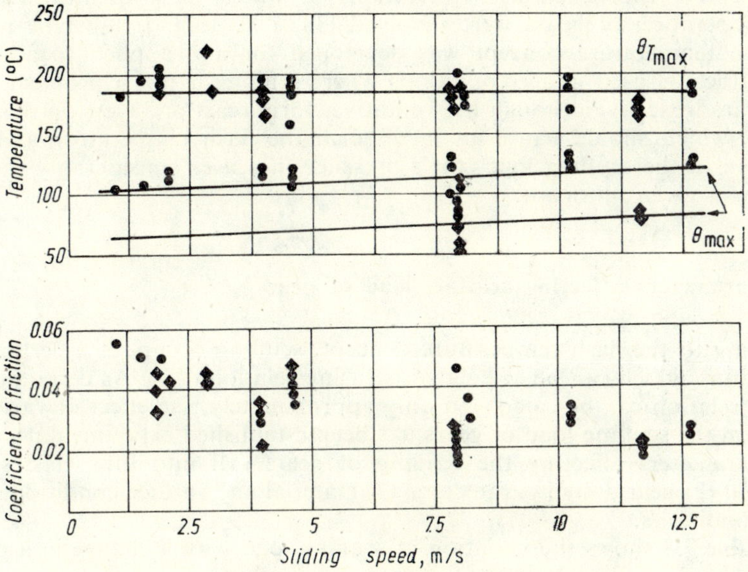

Figure 7.12 Flash temperature θ_{max} and $\theta_{T max}$ at the moment of scuffing and coefficient of friction versus sliding speed for a nonadditivated oil [107]. Lubricant temperature: ◆ 15°C; ● 82°C.

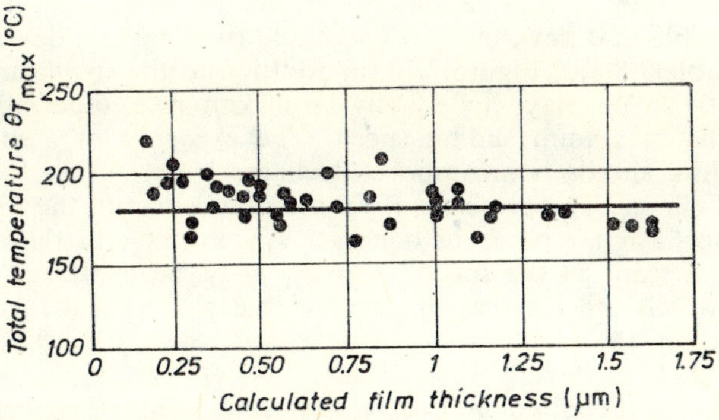

Figure 7.13 Relationship between calculated thickness of the lubricant film and surface temperature at the moment of scuffing [109].

surfaces. Considering the real situation in the stressed area of a pair under concentrated load, the question arises whether it is not possible that the results given in Figures 7.12 and 7.13 are due to the selection of a particular range of the parameters to be investigated.

It has been recently reported that for concentrated load (point or line contacts) the limit of operating without scuffing represents a surface in the three-dimensional space: load, speed, lubricant bulk temperature [198].

To determine scuffing conditions more precisely, other authors have carried out extensive experimental investigations in which most quantities pertaining to the operation of gears were involved [38, 111, 132]. It must be noted that the flash temperature concept was developed for and applied to additive free gear lubricants used under conditions where scuffing is not a problem.

Paradoxically although E.P. additives form reaction layers on the frictional surface, which should cause an increase in the flash temperature and hence a reduction in the scuffing load, their presence increases considerably the scuffing load for a given lubricant.

7.3.2 Parameters affecting scuffing load of gears

According to the flash temperature concept, scuffing occurs at a certain surface temperature which can be calculated by using relation (7.4). As certain quantities for this relation can be calculated only approximately, the effect of various parameters on the scuffing load of gears has been established experimentally [38, 111].

Parameters affecting the scuffing of gears fall into four classes: (1) the lubricant, (2) gear design parameters, (3) material and surface condition, (4) operating conditions.

Table 7.8 shows the variation of scuffing load with a change in a particular parameter when all other parameters are constant.

An important conclusion from the data given in Table 7.8 is the significant effect, comparable with that of the other parameters, of extreme pressure additives on scuffing load. When all other possibilities have been exhausted, the use of extreme pressure additives increases the scuffing load by the ratio 1: 5. As the mechanism of action of these additives has not yet been completely elucidated, it appears possible that new, more efficient additives may be developed.

From Table 7.8 and Figure 7.14, in addition to the significant effect of E.P. additives, tooth shape plays an equally important role, especially the elements determining the maximum sliding speed. The existence of a minimum in the range of medium speeds is an important characteristic of the "scuffing load — sliding speed" curves. The presence of an increasing arm in the high speed range contradicts the flash temperature concept. An increase in the critical temperature of the lubricant as the speed increases may be explained by the short duration for which the lubricant layer is subjected to load. Under a loading time shorter than 10^{-5} s the viscoelastic characteristics of the lubricant film predominate [38].

The diagram presented in Figure 7.15 shows that this behaviour is due to the structure and composition of the lubricant. Not all lubricants have the rising arm characterizing high speeds and for those exhibiting this portion it has different lengths.

Instead of the flash temperature Lechner and Niemann proposed the use of the permanent temperature of the tooth flank which they found to be approximately equal to the temperature in the centre of the tooth [132]. Following numerous experimental determinations they proposed the following relation or the temperature in the centre of the tooth T_m

$$T_m = T_M + Av^{1/3}W^{1/2}, \tag{7.7}$$

where

T_M is temperature for idle running (°C),
W — load per unit length (da N/cm),
v — speed at the rolling circle (m/s).

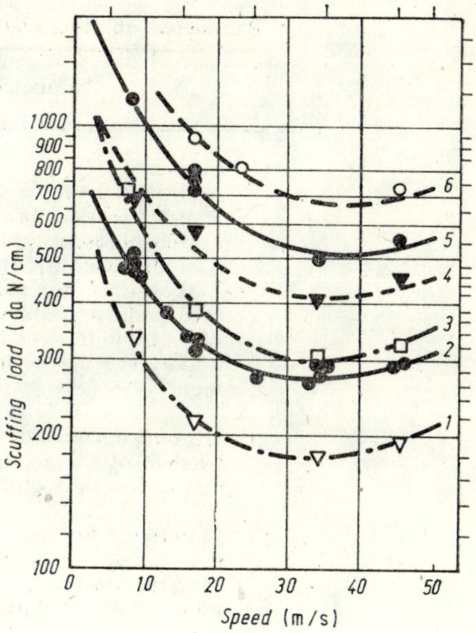

Figure 7.14 Scuffing load versus speed on the rolling circle for various gear characteristics [111].

No.	m	α^0	V_a/V_T	Z_1Z_2
1	6	20	0.754	12:18
2	4.5	20	0.67	16:24
3	6	20	0.67	12:18
4	4.5	20	0.57	16:24
5	4.5	20	0.44	15:24
6	4.0	30	0.45	16;24

where

m ÷ module
α ÷ gearing angle
$V_a \div (V_1 - V_2)$
$V_T \div (V_1 + V_2)$
Z — number of teeth

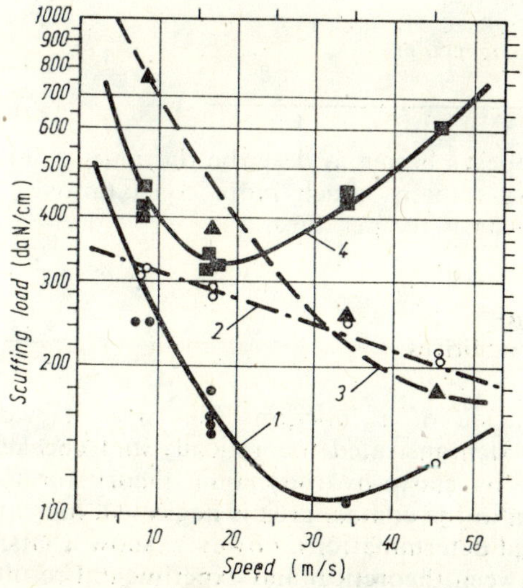

Figure 7.15 Scuffing load versus speed on the rolling circle for various lubricants (gear no. 2, Figure 7.14).

Lubricant: *1*, clean mineral oil; *2*, polyglycol; *3*, additivated turbine oil; *4*, additivated gear oils.

The temperature for idle running is determined by the lubricant temperature while the coefficient A is determined in terms of gear efficiency ε by the relation

$$A = 10\varepsilon, \tag{7.8}$$

where ε is taken in unit fractions.

Table 7.8

Parameters affecting scuffing load of gears (after [111])

Parameter	Scuffing load range
(1) Lubricant	
Additivated (at the same viscosity of the basic lubricant)	1: 5
Lubricant viscosity. E.g. doubling of the viscosity at 50°C:	
a) straight mineral oil	1: 1.5
b) additive containing mineral oil	1: 1.15
Decrease in the coefficient of friction by 1: 2 by the use of a synthetic oil	1: 2.5
2) Gear characteristics	
Change of the geometry	1: 6
Correction of toothing	1: 1.5–2
3) Material – Surface	
Roughness. E.g. reduction by 1:10	1 :2
Material: increase in hardness	1: 4
Phosphating of flanks	1: 1.4
Nitriding	1: 1.8
4) Operating conditions	
Sliding speed, for usual oils	1: 2.5
Sliding speed, for highly refined oils	1: 8
Oil temperature, e.g. a reduction by 20°C	1: 1.8

The calculated temperature T_m appears better to describe the anti-scuffing properties of a lubricant. This quantity remains much more constant over a wider range of sliding speeds than the flash temperature.

7.3.3 The mechanism of scuffing of gears and the role of extreme pressure additives

For concentrated load pairs the existence of a lubricant film over a wide range of operating conditions has been demonstrated theoretically and checked experimentally. The relationships given by elasto-hydrodynamic theory for the lubricant-film thickness when the friction in the contact area is neglected, indicate that the load effect is small. Experimental determinations, however, show a much greater effect of load. The difference between theoretical and experimental results may be significantly minimized if the surface real temperature is used to calculate lubricant viscosity. As this temperature is higher than the bulk temperature, substitution causes a reduction in lubricant film thickness. Measuring the thickness of the lubricant film on a disc machine by means of the discharge technique and the temperature of the surface by means of a sliding thermocouple, O'Donoghue found a satisfactory agreement between the calculated thickness and the measured thickness (Figure 7.16).

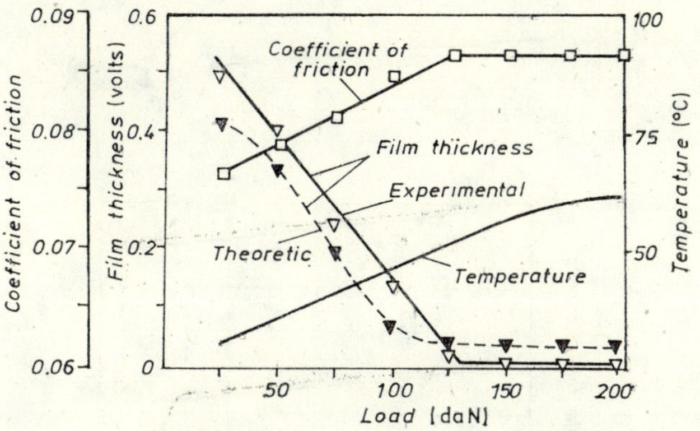

Figure 7.16 Thickness of the lubricant film, surface temperature and coefficient of friction versus load at $\omega = 200$ r.p.m. [57].

Under the conditions shown in Figure 7.16 or 7.17 *a* (low speed) it follows that the scuffing process would occur under a load which is sufficiently high for the thickness of the lubricant film not to be able any longer to ensure the separation of the surfaces of this pair.

The situation is, however, different at high rotational speeds when scuffing occurs even under greater lubricant-film thicknesses (Figure 7.17 *b*). This phenomenon, namely the sudden collapse of the lubricant film even at thicknesses which are sufficient to separate the surfaces indicate the existence of a critical quantity the exceeding of which determines the onset of scuffing.

Cameron identified this quantity with the maximum flash temperature of the surfaces [47]. According to this assumption, even in the presence of a sufficiently thick lubricant film scuffing will occur as a consequence of the accidental contacts between asperity peaks after the surfaces reach the critical temperature.

The results shown in Figure 7.17 and the above conclusions were derived for additive free lubricants. E.P. additives in the lubricant also act by forming a reaction layer on friction surfaces. The formation of a reaction layer has been established directly on gears by using a labelled additive. After running a certain time under a load below the scuffing load, the radioactivity of the tooth flank was determined by autoradiography [37]. Considering the activity to be directly proportional to layer thickness, the layer distribution on the flank surface could be found (Figure 7.18). The maximum thickness coincides with the areas of maximum sliding, where the surface temperature is also high. The existence of such a layer is the result of an equilibrium between the rates of layer formation and wear. In the absence of a detailed interpretation of the role of this layer in preventing scuffing, it is generally agreed that it ensures separation of the metallic surfaces even after the lubricant film has been removed. Concerning the occurrence of scuffing in the presence of a lubricant film it would seem that the reaction layer prevents the onset of scuffing by accidental adhesion.

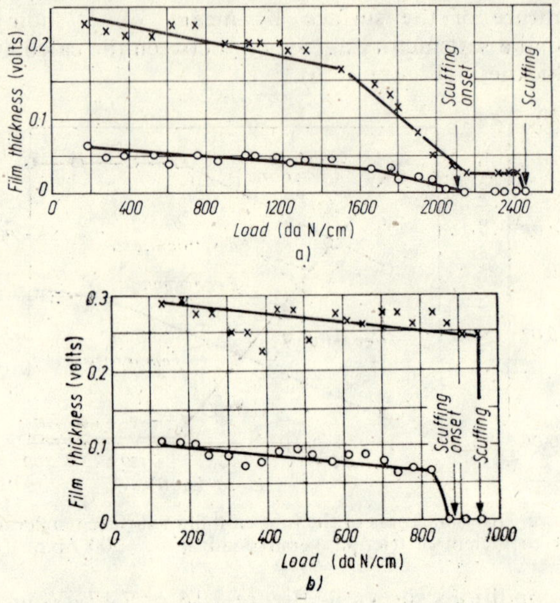

Figure 7.17 Thickness of the lubricant film versus load at low speed (*a*) and high speed (*b*) [47]: × – film thickness on the pitch circle; o – film thickness at tooth tip.

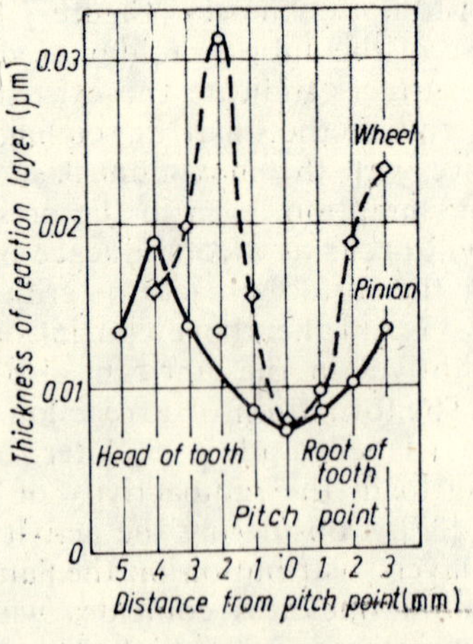

Figure 7.18 Thickness of reaction layer on tooth flank [37].

In the presence of an extreme pressure additive the temperature necessary for the scuffing becomes much higher. The adsorbed layers, which have a much lower desorption point, cannot effectively prevent scuffing. As reaction layers are generally inorganic metallic compounds they have a higher melting point and greater mechanical strength than adsorbed layers. Also, reaction layers may generate a surface which would bind more strongly the adsorbed additive layer.

7.4 Use of additives in the production of lubricants

The use of various types of additives is almost general practice in the present-day production of lubricants. Additive free mineral oils are only used for a limited number of applications where the mechanical, thermal or other stesses are so small that the use of additives is not economically justified. The present tendency to improve the performance of machines and mechanisms especially by increasing specific stesses and reducing weight continuously requires lubricants of improved quality.

The major raw material currently used to produce lubricants is mineral oil which is obtained by processing crude oil. Sometimes, for instance when the working temperature is high synthetic materials are used instead of mineral oils.

The choice of the mineral oil used in the production of a given lubricant depends on cost, viscosity, degree of refining, temperature stability and compatibility with the other ingredients used, or the materials with which it will come into contact. Special attention should be given to the degree of refining of the mineral oil as this significantly affects both cost and the effectiveness of some extreme pressure, antioxidant and other additives [162, 36]. Optimization should take into account the cost of the lubricant.

The operational properties of mineral oils are improved by the addition of several additive substances in various concentrations depending on the nature of the additive and the field of application for which the lubricant is required. For instance, antifoaming additives are used in concentrations of the order of 10^{-5}, while extreme pressure additives may have a concentration of up to 10% or even more.

Depending on their effect the additives used for lubricants may be classified as:

— Pour point depressants used to improve behaviour at low temperatures.

— Viscosity index improvers used especially for multigrade lubricants.

— Antioxidant additives used in lubricants for internal-combustion engines, gears, etc.

— Corrosion inhibitors used in lubricants for turbines, internal-combustion engines, hydraulic oils, etc.

— Detergent — dispersant additives used in lubricants for internal-combustion engines.

— Additives used to improve the friction, wear and E.P. characteristics. They are used in gear lubricants, motor oils, in some hydraulic oils and in cutting fluids.

— Antifoaming additives.

The class of additives used to improve friction, wear and E.P. characteristics includes substances which reduce the coefficient of friction, the wear of surfaces or increase the scuffing (seizure) limit. In the literature, various other less precise names exist for additive classification such as oiliness additives, additives used to improve the film resistance, etc. Because of their ambiguity such names are not recommended for use.

As a consequence of the vast number of lubricants produced, a classification of these is required. This classification, usually made according to the field of application, is given as:

— motor oils,
— gear and transmission oils,
— turbine and hydraulic oils,
— metal working oils,
— oils for other uses (e.g. anticorrosive oils).

All these classes have subclasses such as the class of motor oils. This comprises: gasoline motor oils, automotive diesel oils, naval oils, railway diesel engine oils, etc.

For all oils used with machinery operating under mixed or boundary lubrication conditions even only intermittently, additives should be used to improve their frictional and wear characteristics.

A recognized application of these additives is for general gear lubricants and especially for automotive gear lubricants. The development of extreme pressure additives is related to the introduction of the hypoid gear in vehicle transmissions. Gradually, due to specific transmitted power increases, use of this class of additives has been extended to the lubricants for gearboxes and other heavily stessed gears.

At present, additives are used to reduce friction or wear in other applications such as oils for machine-tool guides and hydraulic drive oils. Sometimes, the role of antiwear or E.P. additive is taken over by certain multifunctional additives such as Zn-dialkyldithiophosphate. This additive, in addition to its antioxidant effect, also reduces wear in camtappet assemblies. Extreme pressure additives are also used in metal working oils. The controlled reactive effects and anti-scuffing effects of certain additives are used in running-in lubricants.

The use of additives to improve the frictional and wear characteristics of lubricants may have a wide range of application as the lubricant is an element which, in addition to the materials of the pair, contributes to improved machine performance. Formulation of a new and effective lubricant is difficult. It is necessary to take into account the physico-chemical properties of the oils and additives used and the operating parameters of the machinery to be lubricated.

7.4.1 **Gear lubricants**

Gear lubricants are an important class of lubricants which utilize extreme pressure additives and additives for the reduction of the coefficient of friction. From the point of view of formulation, the properties of gear lubricant may be classified in two groups: inherent properties and desired properties. The first group involves: viscosity, flash point, etc., namely quantities which are inherent in the base mineral oil. The second group includes the desirable properties of the formulated lubricant: friction and wear behaviour, anti-scuffing properties, viscosity-temperature relationship, resistance to heat, corrosiveness, etc. The use of additives may partly provide the desired properties.

Depending on the method of test, the properties of the lubricants are classified into physical and chemical properties.

Compared with the number of standard physico-chemical tests (about 60 FTM Std.) the number of tests of anti-wear and anti-scuffing properties is much more reduced. Standardization of techniques for the determination of these properties has encountered several difficulties due to the great variety of the operating conditions in machines and to an incomplete knowledge of the phenomenon of mixed and boundary lubrication. A list of FTM for the load carrying capacity of gear oils is given in Table 7.9.

Table 7.9

Federal test method for load carrying capacity of gear oils (FTM 791)

Method	FTM No.
Load carrying capacity:	
– S.A.E. test machine	6501.1
– Mean Herz load (four-ball machine)	6503
– of universal gear lubricants by the Timken machine	6505
Load carrying:	
– ability of lubricating oils (Ryder Gear Machine)	6508
– ability of lubricating oils at 400°F	6511
– wear, stability and corrosion characteristics of universal gear lubricants in axles under conditions of high torque and low speed	5317
– characteristics of universal gear lubricants in axles under conditions of high speed	6504
– wear and E.P. characteristics of gear lubricants in axles under conditions of high speed-low torque operation, followed by low speed-high torque operation	6506
– and E.P. characteristics of gear lubricants in axles under conditions of high speed and shock loading	6507
Navy gear wear test	335

Despite difficulties and objections, a large number of machines and methods have been devised to investigate the friction and wear properties of lubricants and additives. Although many of these methods have proved to be valuable for basic or applied research, especially for the design of a new lubricant or additive, the results obtained on such apparatus are not generally comparable and can be extrapolated only partly for practical applications. As the theory of mixed and boundary lubrication improves and the mechanism of action of lubricants and scuffing phenomenon are better understood, standard test methods will be evolved. More and more countries have now included in their standards several laboratory methods for lubricant tests, such as the four-ball machine, the Timken machine, and the gear test machine.

Classification. Specifications

To define a lubricant, so-called "specifications" are used. Specifications, given in the form of a list summarising the physico-chemical conditions a lubricant should fulfil, together with the method of production to be used ensure a quality product. Most of the conditions included in specification have been established empirically, following formulation of a lubricant which yielded good results.

Vehicle transmission oils. The S.A.E. classification of these lubricants as a function of viscosity is given in Table 7.10.

Table 7.10

Classification of vehicle transmission oils

SAE Class	Viscosity at – 17.8°C				Viscosity at 98.9°C			
	cSt		°E		cSt		°E	
	min.	max.	min.	max.	min.	max.	min.	max.
75	—	3557	—	430				
80	3257(*)	21716	430(*)	2867	4.18	—	1.33	—
90					14.24	25.0(**)	2.249	3.467
140					25.0	42.7	3.467	5.707
250					42.7	—	5.707	—

(*) Minimum viscosity at – 17.8°C is cancelled as a limit value if the viscosity at 98.9°C is higher than 6.66 cSt (1.536°E).

(**) Maximum viscosity at 98.9°C is cancelled as a limit value if the viscosity at – 17.9°C is lower than 16 2900 cSt (21.500°E).

To characterize the level of additives which is a very important element in judging the performance of a lubricant the API (American Petroleum Institute) suggested the following six groups:

API–GL–1. Designates the type of service characteristic of automotive spiral-bevel and gear axles and some manually operated transmissions operating under such mild conditions of low unit pressures that straight mineral oil can be used.

API–GL–2. Designates the type of service characteristic of automotive type worm-gear axles operating under such conditions of load, temperature and sliding velocities that lubricants satisfactory for API–GL–1 service will not suffice.

API–GL–3. Designates the type of service characteristic of manual transmissions and spiral-bevel axles operating under moderately severe conditions of speed and load.

API–GL–4. Designates the type of service characteristic of gears particularly hypoid in passenger cars and other automotive type equipment operated under high-speed, low-torque, and low-speed, high torque conditions. Lubricants suitable for this service are those which provide anti-scuffing protection equal or better than that of a reference oil (CRC reference gear oil RGO 105).

API—GL—5. Designates the type of service characteristic of gears, particularly hypoid, in passenger cars and other automotive equipment, operated under high-speed, shock-load; high-speed, low-torque, and low-speed, high-torque conditions (CRC reference oil RGO 110).

API—GL—6. Designates the type of service characteristic of gears, specifically high offset hypoid gear (above 50.8 mm and approaching 25% of ring gear diameter) in passenger cars and other automotive equipment operated under high-speed, high-performance conditions (CRC reference gear oil L—1000).

According to an older classification, there were three major classes of gear lubricants: regular, mild E.P. gear lubricants and E.P. gear lubricants. The first type would be used for light loads, the second especially for vehicle transmissions, while the last for hypoid gears and for heavily loaded straight gears.

It is obvious that the load capacity of a certain gear lubricant cannot be defined only by physico-chemical tests. Specifications should also include determinations concerning friction properties, wear and resistance to scuffing. A first step to overcome these difficulties is to introduce new tests based on laboratory apparatus and methods of lubricant testing. However, as the test conditions in such apparatus sometimes differ substantially from the operating conditions in real gears, several authors and specifications require experiments to be carried out with a full scale test rig.

Industrial oils. This class includes the oils used for industrial equipment operating under conditions of moderate loading. Although in many cases additive free oils are used in industrial installations, there is a tendency to replace these by lubricants containing some additives, especially under conditions of heavy loading. The AGMA 250.02 specifications divide industrial gear oils into 11 types (Table 7.11).

Table 7.11

Viscosity of industrial gear oils according to AGMA 250.02 standards. Viscosity range in cSt

	AGMA Lubricant No.										
	1	2	3	4	5	6	7	7 comp.(*)	8	8 comp.(*)	8A(**) comp.(*)
Viscosity at 37.8°C (100°F)	39—52	61—78	106—152	152—217	—	—	—	—	—	—	—
Viscosity at 98.9°C 210°F	—	—	—	—	16—22	22—26	26—32	26—32	32—61	32—41	41—54

(*) Additive fortified to reduce the coefficient of friction.
(**) The table does not include oils nos 9, 10 and 11.

Choice of the lubricant

The choice of gear lubricants, which is a most important problem, should be made according to their properties and the characteristics of the gear to be lubricated. However, owing to the great complexity of the friction process in gears, at present the choice of a lubricant is made empirically. For new types of gears the lubricant is often adapted following exhaustive tests.

Elasto-hydrodynamic theory permits, if the real surface temperature is taken into account, calculation of the minimum lubricant film thickness as a function of the lubricant viscosity and the principal mechanical characteristics of the gear such as speed, radius of curvature and material.

By choosing a minimum permissible lubricant film thickness the required viscosity is obtained. Two difficulties are, however, encountered with this method. The first is in the choice of a minimum film thickness as it is known that scuffing may also occur with film thicknesses exceeding the asperity level, and that in the presence of several extreme pressure additives the calculated thickness may be smaller than the mean asperity height although the pair is not adversely affected in operation [47, 109]. The second difficulty is to calculate the gear tooth surface temperature, a quantity which is determined by the gear heat balance.

To consider the scuffing phenomenon which becomes important at high sliding speeds or high temperatures the criterion of critical temperature or duration temperature of the tooth flank is used. The use of these methods assumes that the critical temperature (flank duration temperature) of the lubricant used is known. This quantity, however, should be determined experimentally.

Niemann and Lechner [110] proposed a method which considers both the physico-chemical properties of the lubricant and the nature of the gear material. In this method the maximum moment M_i which can be endured by the lubricant tested is determined on a F.Z.G.-type gear installation (determination A/8.3/90). With this value, corrected for the case when speeds other than 8.3 m/s are used, the permissible load in the lubricant film (K_F) is calculated. Using the design parameters of the gear which is to be lubricated, the effective stress (K_W) is then determined. For a safe, scuffing-free operation the ratio

$$S_F = \frac{K_F}{K_W}, \tag{7.9}$$

termed the scuffing safety coefficient should fulfil the condition $S_F \geqslant 3$. If S_F comes out smaller, a lubricant should be selected which provides a greater value for M_i. That is, the lubricant should be selected on the basis of the values of M_i determined experimentally.

Considering the uncertainties concerning the calculation of the required lubricant viscosity and especially of the additive content, empirical recommendations are followed. These take into consideration the current characteristics of gears, materials and processing techniques.

In industrial gears where the danger of scuffing is small, an additive free lubricant may be used and the viscosity depends on the transmitted power and pinion speed. Tables 7.12 and 7.13 give recommended viscosities for industrial transmissions according to DIN 51509.

AGMA Recommendations No 250,00 also take into account the environmental temperature (Table 7.14).

Table 7.12

Viscosity in cSt at 50°C for cylindrical or bevel industrial gears (after DIN 51.509)

Power (HP) \ Pinion speed r.p.m	100 ÷ 250	250 ÷ 750	750 ÷ 2500	>2500
1–5	68±6	49±5	36±4	36±4
5–100	68±6	68±5	49±5	36±4
> 100	92±7	68±5	49±5	36±4

Table 7.13

Viscosity in cSt at 50°C for worm gears (after DIN 51.509)

Power (HP) \ Worm speed r.p.m	100 ÷ 250	250 ÷ 750	750 ÷ 2500	<2500
1– 20	114±8	92±7	68±6	49±4
200–100	25 ÷ 45	114±8	92±7	68±6
>100	cSt/100°C	25 ÷ 45 cSt/100°C	114±8	92±7

With all recommendations the lubricant viscosity decreases as speed increases and increases as the power increases. This is to be expected considering the relationship between viscosity and the thickness of the lubricant film between the tooth flanks.

As an approximate criterion for the change to additive fortified lubricants (with E.P. and anti-wear additives) Bartz [22] recommends the following stress levels. For quenched gear teeth a pressure of 80 kg/mm^2 calculated on the rolling circle and for unquenched teeth a pressure of 40 kg/mm^2.

For heavily stressed gears operating at high sliding speeds the use of extreme pressure and anti-wear additives is essential for designs in current use. Lubricant type and additive formulation are finally established by tests. For any critical transmission the lubricant used is usually selected following collaborative discussions between the gear designer and lubricant supplier. Such is the case with vehicle and turbine gears, etc. Several recommendations follow such collaboration which in certain instances may also involve other types of gears.

Table 7.14

Viscosity of lubricant for cylindrical and bevel industrial gears (after AGMA No 250.02)

Gear type	Gear size (distance between axes for low velocity stage)	Lubricant No. in AGMA (*) classification for an environment temperature of: −9.4 ÷ 15.6°C	10 ÷ 52°C
Cylindrical gears (single reduction)	< 200 mm	2	3
	between 200 mm and 500 mm	2	4
	> 500	3	4
Cylindrical gears (double reduction)	< 200	2	3
	between 200 mm and 500 mm	3	4
	> 500	3	4
Cylindrical gears (triple reduction)	< 200	2	3
	between 200 and 500 mm	3	4
	> 500 mm	4	5
Planetary gears	outer size of casing		
	< 400 mm	2	3
	> 400 mm	3	4
Straight or spiral bevel gears	gear outer diameter		
	< 300 mm	2	4
	> 300 mm	3	5
High speed gears(**)	all sizes	1	2

(*) See Table 7.12.
(**) The pour point should be below the minimum temperature of the environment.

7.4.2 **Cutting fluids**

The fluids used in metal cutting should possess the following properties: high specific heat, good frictional and wear characteristics under conditions of extreme pressure and be non-corrosive.

The cutting process is characterized by high stress, great deformation and high contact temperature. Heat is produced by friction between the tool and the workpiece, the tool and the chip and by the plastic deformation to which the chip is subjected. At high cutting speeds, such as in metal turning, temperature is the limiting factor and the main role of the fluid used is to cool the working area. The temperature may, however, be lowered by reducing the coefficient of friction between the tool and the workpiece or between the tool and the chip which is effected by the friction modifiers and the extreme pressure additives in the cutting fluid. These additives also reduce material transfer and improve the quality of the worked surface.

Three types of cutting fluids are used:

— Aqueous solutions containing anti-rust additives (surface-active agents). These liquids have a high specific heat but have poor lubricating properties.

— Cutting oils containing extreme pressure additives and anticorrosion additives. These blends have good lubricating properties but poor specific heat.

— Emulsions of cutting oil in water which combine the advantages of the other types. An emulsifying agent is required, however, in the base oil.

Sulphur, chlorine and phosphorus compounds are used as E.P. additives. For cutting oils, chemical E.P. additives are sometimes successfully replaced by suspensions of graphite or molybdenum disulphide. E.P. additives used in cutting fluids should possess increased reactivity compared with those used in lubricating oils. Increased reactivity is required owing to the severe conditions under which the cutting process takes place, namely high pressure and possibly high sliding speed. Under these conditions, access of the active substance to the area of maximum stress becomes difficult. The reaction time should be shorter than the rate of generation of the new surface.

The effect of an extreme pressure additive is most obvious in the machining of hard or special alloys. For instance, during threadcutting with taps of a steel quenched to a hardness of HRC 52, the number of threaded holes decreases from 75 if trichlorethane is used as the cutting fluid to 0 in the case of sulphurized oil (Figure 7.19). The difficulty of additive selection increases as the difficulty

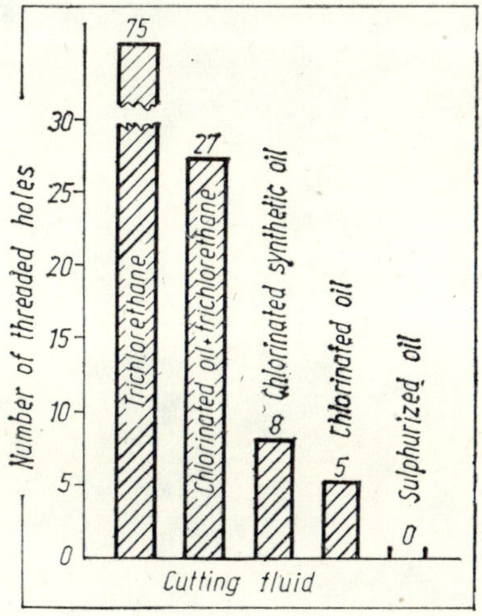

Figure 7.19 Effect of cutting fluid on tap endurance in quenched steel machining [183].

to work the alloy increases while the effect on tool life may be greater. Tool life is important economically especially for automatic machines. An exploratory selection of extreme pressure additives for cutting oils may be made by classical laboratory lubricant and additive testers. Additive effectiveness is confirmed by direct machining tests to determine tool wear and the quality of the surface finish.

[illegible] These plants have good [illegible] [illegible]

[illegible] chlorine and [illegible]

[illegible] The effect of [illegible]

Figure [illegible]

[illegible] work the alloy [illegible] [illegible]

8 Experimental methods for the study of the anti-wear and anti-scuffing properties of additives and lubricants

Numerous experimental devices have been created for the study of the anti-wear and anti-scuffing properties of additives and lubricants. These devices and methods are used either for basic research or as test methods. Many are used only to compare the properties of lubricants and materials.

Devices used for the determination of anti-scuffing and anti-wear properties fall into two principal classes:

(a) laboratory test devices which use a relatively simple test specimen configuration to model frictional processes occurring in real machines or perform a frictional process under specific conditions,

(b) apparatus which uses machine parts and is adequately equipped with measuring instruments.

The last stage of formulation of a lubricant is testing under real service conditions. These tests are expensive and of long standing.

Investigation costs increase substantially when passing from laboratory tests to real service tests. This is why the laboratory stage is used although many authors consider that many laboratory devices with a simple pair as currently used do not reproduce faithfully the conditions existing in real machines.

For basic research, however, laboratory devices are essential as they permit rigorous control of operating parameters.

8.1 Classification of laboratory test devices

Owing to different reasons such as simplicity, advantageous design solutions and the modelling of certain machine parts numerous laboratory devices have been designed.

In order to define the field of application and the possible uses of such devices a classification has been proposed. Some authors have confined themselves to summing and describing the devices [29]. Although this is only an initial step such publications are useful especially as many authors do not specify exactly the characteristics of the devices they have used.

Other authors have endeavoured to classify existing devices according to the criterion of the type of contact between the mating parts [43, 136]. Using such a criterion, Pavelescu has classified most existing devices [136].

A classification may be made according to various criteria. It is useful, however, if these are the main parameters of friction and wear processes. Unfortunately the establishment of these parameters is closely related to the establishment of the laws of friction and wear processes which have not been satisfactorily accomplished especially for mixed lubrication conditions. However, based on existing knowledge several other quantities may be also selected as classification criteria. These include the feasible pressure and temperature ranges and their distribution and the sliding/rolling speed. A classification based on these criteria would produce comparable classes of devices. Then, it would be expected that sometimes the results obtained with two different devices would not be comparable even for relative results.

The above parameters would be useful for a description of the characteristics of the test device. This could be accomplished simply and precisely by indicating the range of possible values for the basic quantities obtainable by means of the test device. Indications concerning the accuracy and stability of the values may be useful. An accurate determination of the operating conditions is of primary importance for comparing results obtained on different devices with those obtained on real machines.

8.1.1 **The four-ball machine**

In 1933 Boerlage proposed the four-ball machine for determining lubricant quality. The friction pair consists of three stationary balls pressed against a rotating

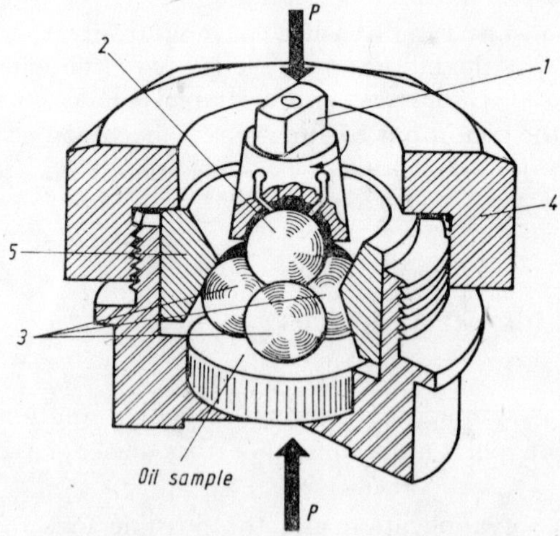

Figure 8.1 Main elements of a f our-ball machine.

ball. The quality of the lubricant was established by the size of the wear scar or the seizure load. Due to its simplicity and use of a simple pair, this device has enjoyed wide popularity. Several recent systematic studies have clarified its possibilities and defects [122, 156, 88]. The four-ball machine is included in several standards and specifications: Socialist Republic of Romania State Standard 8618-70; FTM no. 791 a/6503; ASTM D 2596-67, DIN 51350.

The four-ball machine (Figure 8.1) consists essentially of a vertical driving shaft provided with a conical device (*1*) at its lower end to hold the moving ball (*2*) which is pressed against three stationary balls (*3*) fixed by a nut (*4*) and a conical part (5). The stationary ballholder is mounted on an axial bearing so that it can rotate and displace in the vertical direction. The load is applied on the stationary balls by a lever device. The friction moment acting on the fixed stationary balls by the rotation of the moving ball (*2*) is transmitted by means of a lever to a measuring device.

The lubricant to be tested is introduced into the cup holding the stationary balls. Balls of 12.7 mm (1/2") diameter are usually used. These are especially processed to ensure a high dimensional accuracy as well as uniform hardness and surface quality. The speed depending on the type of device used varies between 1425 and 1770 r.p.m. The load is applied progressively and discontinuously, the load interval depending on the measuring method used. Either the wear spot is measured on the three stationary balls after a fixed running time or the load is established at which seizure or welding of the pair occurs.

On the basis of these measurements the following quantities may be determined: the Mean Hertz Load, the maximum load without seizure or the initial seizure load, the maximum load without seizure for a running time of 2.5 s and the welding load.

By plotting the wear spot diameter in terms of the applied load on a log. log. diagram a curve as shown in Figure 8.2 is obtained. Some methods choose the load intervals so that log P increases for each interval by 0.05. The portion *AB* illustrated in Figure 8.2 represents the portion of wear without seizure; *BC* is the initial seizure portion; *CD* is the immediate seizure portion; *D* is the welding load.

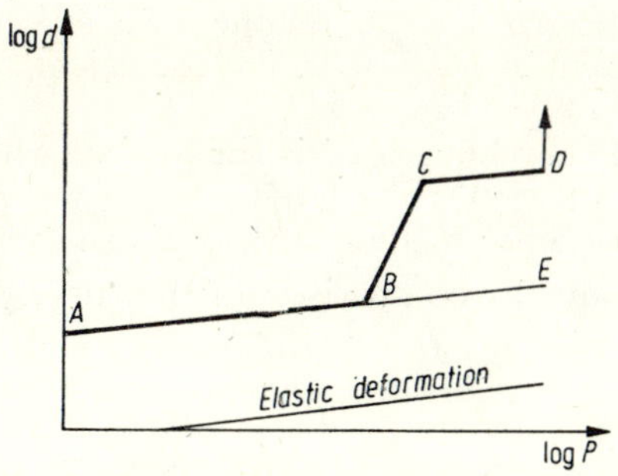

Figure 8.2 Variation of the wear spot diameter with the applied load: *ABE*, compensation line; *B*, maximum load without seizure; *BC*, seizure onset region; *CD*, immediate seizure region; *D*, welding load.

The Mean Hertz Load is determined by measuring the wear scars for 10 seconds running time under increasing loads until welding occurs. Tests begin with a load of 40 daN. If the welding load is lower or equal to 316 daN, the number of runs is completed to 20, also carrying out runs under loads lighter than 40 daN. If the welding load exceeds 316 daN, then 20 runs are selected commencing with the welding load.

Denote by A the sum

$$A = \frac{1}{m} \sum_{1}^{m} P_i \left(\frac{D_h}{\overline{D_u}} \right)_i , \tag{8.1}$$

where $P_m = 316$ daN or the welding load if this is less than 316 daN, and by B the sum

$$B = \frac{1}{n} \sum_{1}^{n} P_i \left(\frac{D_h}{\overline{D_u}} \right)_i , \tag{8.2}$$

where

P_n is the welding load if this is greater than 316 daN and

D_h — elastic contact spot diameter for steel balls of 12.7 mm diameter; $D_h = 8.73 \cdot 10^{-2} P^{1/3}$,

$\overline{D_u}$ — mean wear spot diameter corresponding to the load i, then the load-wear factor M.H.L. for $m + n = 20$ is given by the relation

$$\text{M.H.L.} = \frac{A + B}{20} . \tag{8.3}$$

This quantity is considered to have reasonable reproducibility and allows lubricant selection according to lubricating capacity under extreme pressure conditions [88, 189].

The initial seizure load is usually determined from the intersection of the portions AB and BC on the wear-load curve or directly as the maximum load for which no increase in the coefficient of friction occurs for 60 seconds running time.

The welding load is the load at which three successive weldings of the pair occur after 60 ± 1 s [185].

Another method of assessing lubricant qualities by means of the four-ball machine is by comparison of the curves log $\overline{D_u}$ in terms of log t [122].

Although the reproducibility and repeatability of the above quantities are within acceptable limits, the most useful results are obtained by M.H.L. However, it is considered that the results cannot be correlated satisfactorily with results from gears. The four-ball machine may be used successfully to control the additive content of a given lubricant or for basic research on the mechanism additive action [190].

The contact surface temperature is an important parameter in the behaviour of E.P. anti-wear additives. With the four-ball machine, this parameter is determined both by calculation and experimentally by the incorporation of a thermocouple [193].

8.1.2 Line contact test machines

The Timken machine uses a friction pair consisting of a rotating bush (*1*) pressed against a rectangular block (*2*). The loading device, provided with a knife hinge also permits measurement of the coefficient of friction (Figure 8.3). The maximum load without seizure is determined for a 10 min. running time. Wear of the fixed part may also be determined by measurement of the wear scar with a Brinell microscope.

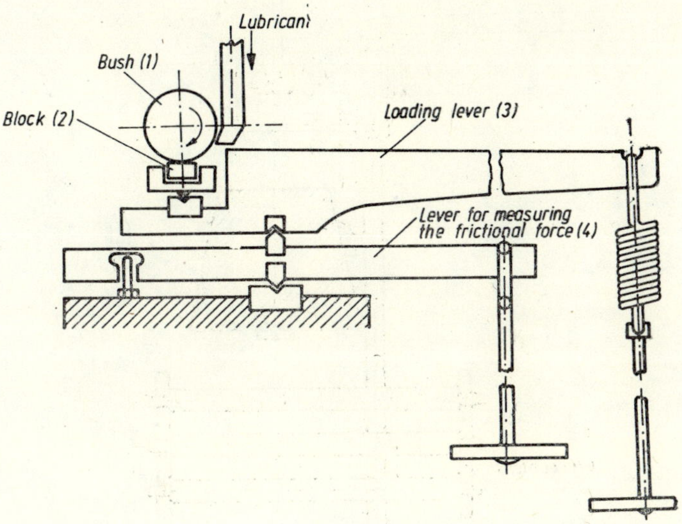

Figure 8.3 Schematic diagram of a line contact machine (Timken).

The coefficient of friction may be measured by balancing out the lever (*4*) in the 10th minute of running. During the run the temperature of the lubricant should be 125°F [188]. The elements of the pair are made of steel, quenched to a hardness of 58–62 HRC.

Although simple and less flexible, the Timken machine is used in several specifications for extreme pressure lubricants.

The I.M.A. machine [134, 135] also uses a friction pair consisting of a bush (*1*) in contact with a plate (*2*) (Figure 8.4). The loading force is applied to the plate through a lever (*3*). To ensure uniform load distribution over the whole contact band the lever and the plate are connected by a hinge. The pairs are totally immersed in the lubricant to be tested.

To determine bush wear a radioactive tracer is incorporated in the bush material. The material separated from the bush as a consequence of wear is collected in the oil container (*4*) and measured radiometrically by means of the counter group (*5*). The protecting screen (*6*) which may be rotated to lower the radioactive oil container, isolates the counter system (*5*) from the radiation emitted by the radioactive bush (*1*). For high activities the transducer system may be isolated in a screened enclosure.

Integral measurement of the lubricant volume, as used in the I.M.A. machine, eliminates errors due to the non-homogeneous distribution of wear debris in the oil volume and by sedimentation of wear debris. These sources of errors are characteristic of the measuring method using oil samples or oil circulation.

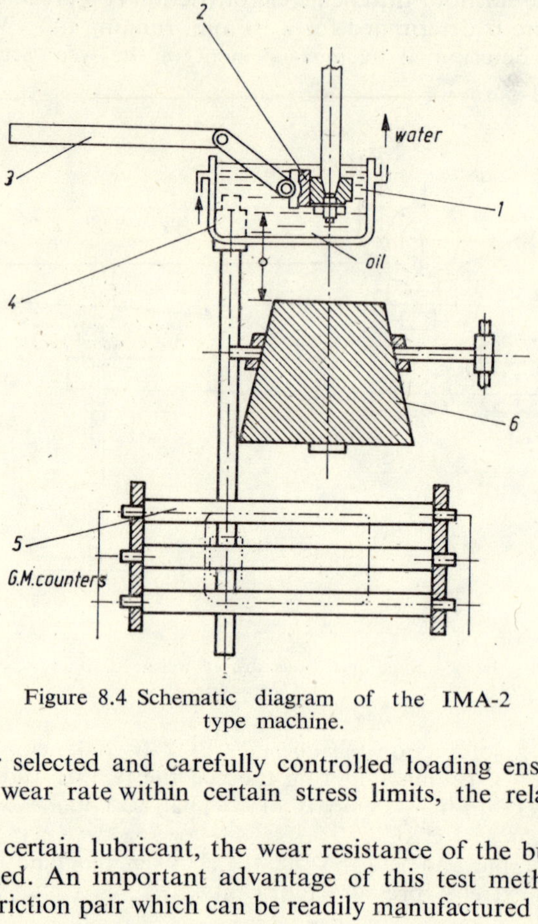

Figure 8.4 Schematic diagram of the IMA-2 type machine.

The type of pair selected and carefully controlled loading ensures good reproducibility of the wear rate within certain stress limits, the relative error being of $\pm 8\%$.

By using a certain lubricant, the wear resistance of the bush material may also be determined. An important advantage of this test method is the simple geometry of the friction pair which can be readily manufactured from the material to be studied.

8.1.3 **Rolling-disk machine**

This type of device allows simultaneous sliding and rolling motion. These two types of motion are characteristic of gears with the exception of worm gear. Selection of the roll diameters, loads and speeds of the two rolls enables the running conditions corresponding to any point on the tooth flank to be simulated. As distinguished from the case of a gear, these conditions are kept constant which is considered to be detrimental for the reproduction of the conditions existing in gears, but advantageous for basic research. The geometrical simplicity of the specimen pair allows improved dimensional accuracy, well-controlled surface treatment and quality.

A rolling-disk machine consists essentially of two discs, of which one has a moving axle to achieve the normal force (*1* and *2*, Figure 8.5). The discs

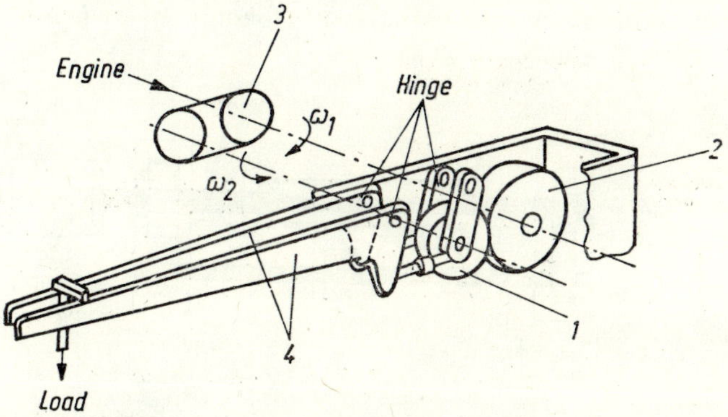

Figure 8.5 Rolling-disc machine.

are driven in opposite directions at different speeds by the driving system (*3*).

Loading is achieved by two independently hinged levers (*4*) which press the moving disc (*1*) against the other disc (*2*), the contact taking place in the direction of the generatrix. Parallel alignment of the two discs is important for satisfactory machine operation. The selected design solution should ensure contact along the whole generatrix for any load and speed; a slight inclination of the axes determines a significant pressure rise in the contact area.

For E.H.D. lubrication studies this machine has facilities for the measurement of the pressure in the contact area and lubricant film thickness.

Determination of the load carrying capacity of a lubricant on an SAE type rolling-disc machine expressed as the maximum load without scuffing is a standard test method [187]. After running for 30 min. the load is continually increased until scuffing occurs.

8.1.4 **Gear type lubricant test machines**

Owing to the complexity of gear lubrication several authors prefer test machines using a pair of gears. With this set up the effect of other parameters such as: tooth geometry, processing accuracy, surface quality, and surface treatment may be examined. There is a complicated relationship between these quantities and the characteristics of the lubricant, which cannot always be reproduced on a simple test machine. Thus this type of machine represents an intermediate stage between tests on simple machines and those on real machines. However, because of the complicated geometry of gears, measurement of the basic quantities pertaining to the processes occurring on the tooth flank is difficult to carry out with these machines. Generally, with conventional machines wear and scuffing load are usually measured. Schematically, such a machine (with a closed force circuit) consists of the working gear and engine gear joined by two shafts (Figure 8.6). Loading of the working gear is achieved by the twisting of one of the shafts. Thus, the driving engine has to compensate only for the frictional losses in the gears and bearings. The main disadvantages of this system are the necessity to stop the test in order to change the load and that the clutch spindle may give use to vibrations in the whole assembly.

Tables 8.1 and 8.2 summarize the gear parameters and working conditions for the more common test machines with loading by twisting. These are the

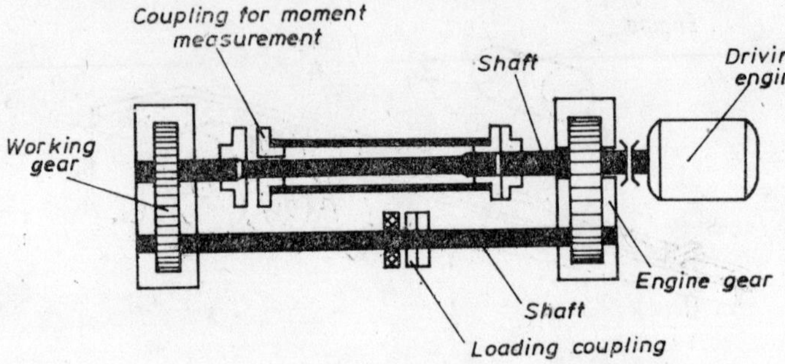

Figure 8.6 Gear-type lubricant test machine.

Table 8.1

Characteristics of gears used in FZG, IAE and Ryder apparatus

	Dimension	FZG A	FZG C	IAE	Ryder
Width between centres	[mm]	91.5	91.5	82.55	88.9
Effective tooth width	[mm]	20	20	4.77	6.35
Module	[mm]	4.5	4.5	5.08	3.175
No. of teeth: — pinion	—	16	16	15	28
— driven wheel	—	24	24	16	28
Addendum modification	—	0.8635 −0.5	+0.1817 +0.1817	0.375 0.494	0
Pressure angle	[°]	20	20	20	22.5
Pressure angle after correction	—	22.5	22.5	26.317	22.5
Speed on the pitch circle (v)	[m/s]	$-0.00383\ n_1$		$0.00417\ n_1$	$0.00465\ n_1$
Maximum sliding speed: — pinion — driven wheel	[m/s]	$0.67\ v$ $0.15\ v$	$0.44\ v$ $0.44\ v$	$0.46\ v$ $0.46\ v$	$0.28\ v$ $0.28\ v$
Load at tooth tip: — pinion — driven wheel	daN/mm^2	$5.96\sqrt{P}$ $4.96\sqrt{P}$	$4.89\sqrt{P}$ $6.96\sqrt{P}$	$10.2\sqrt{P}$ $10.6\sqrt{P}$	$8.9\sqrt{P}$ $8.9\sqrt{P}$

Table 8.2

Main data concerning the FZG, IAE and Ryder methods

	FZG	IAE	Ryder
Pinion speed n_1 (r.p.m)	2250(4500)	2000/4000/6000	10,000(5000.2000)
Lubrication	by immersion	by jet	by jet
Oil temperature, °C	initial 90	60/70/110	74
Oil flow (l/min.)	oil quantity: 2 l	0.285/0.57	0.27
Running time for each load (min.)	15	5	10
Normal load on the tooth for the 1st loading step (daN/cm)	5	82.0	66.0
Increase in the normal load on the tooth for one step (daN/cm)	14.5÷115	41	66
Load carrying capacity defined by	scratches or scuffing and transition to severe wear	(1) scratches or scuffing at tooth head (2) scratches or scuffing at tooth root	deteriorated surface: 22.5 % of flank surface

FZG, IAE and Ryder (loaded hydraulically) machines which are the standard gear lubricant test machine used in countries where they were developed.

The FZG method of normal test, denoted by A/8.3/90 uses tooth characteristics given in Table 8.1, with a pitch circle velocity of 8.3 m/s and an initial oil temperature of 90°C. The load is applied progressively, with approximately 15 min. running time for each load, which corresponds to 21,700 rotations of the engine shaft. After each load step the gear is demounted and weighed to determine wear. The load is increased up to the 12th load step or up to the onset of severe wear as a consequence of scuffing. After scuffing has occurred the number of rotations is completed to 21,700. Wear of the working gear is plotted as a wear curve in terms of load steps or transmitted energy (Figure 8.7). It is considered that severe gear wear had occurred when the increase in the total wear ΔG is greater than 10 mg for a load step (station V on Figure 8.7). The maximum admissible moment M_i determined by the above method which is a characteristic of the lubricant tested may be used for calculation of the scuffing load for various gears and operating conditions.

Besides determination of the wear curve and scuffing load, the FZG method recommends observation and description of the changes on the gea· tooth flanks, such as scratches, ridging and colouring.

8.2 Discussion of results obtained with laboratory test devices

From the point of view of reproducibility and repeatability, lubricant testers are not different to other measuring instruments. The spread of results can be reduced if every test reproduces the same working conditions and by improved knowledge of the process under investigation.

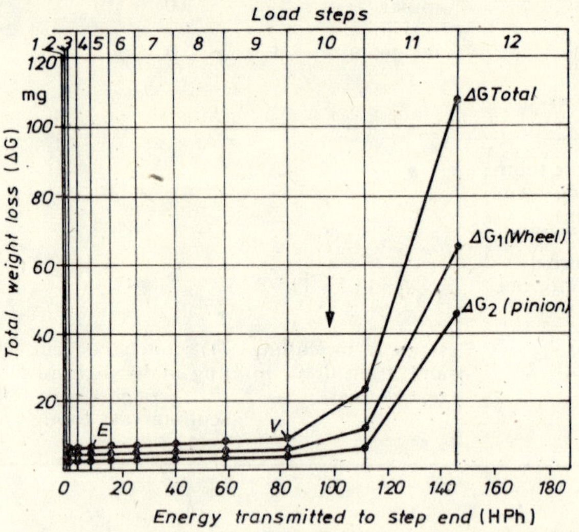

Figure 8.7 FZG lubricant test method results graph.

Another aspect of the use of laboratory test devices is comparison of the results obtained with different devices and with the results from practice. Comparative studies of the same lubricant with different laboratory test devices have shown that the results do not always correlate even if expressed in a relative manner [21, 22]. The results of such a study with six test devices are listed in Table 8.3. The most effective lubricant is denoted by 1 while the least effective is denoted by 5. Intermediate values correspond to some intermediate performance. Although in Table 8.3 a clear discrepancy exists between additive containing and additive free oils, there are also cases where a lubricant is denoted differently from 1 through 4 or even 5. Such a large variation arises from the different test conditions occurring in the devices. It is also concluded that discrepancy in the performance of two lubricants also depends on the load and speed ranges used.

The lack of correlation between results obtained with various laboratory test devices is not considered to detract from their use. A similar lack of agreement between results is also found with full scale tests for which reproducibility of the results is poorer and the cost of testing is higher.

Table 8.3

Relative quality of several lubricant determined by some laboratory test devices (after [21])

		Running in oils		Hypoid oils				Universal oils		Non-additivated oils	
		A	B	C	D	E	F	G	H	I	J
Scuffing safety	Timken (scuffing load)	1	2–3	3	2	4	2	2	3	5	5
	Four-ball lubricant tester (welding load)	1	1	1–2	4	2–3	3	3–4	3	5	5
	Four-ball lubricant tester (seizure load at 2.5 s)	1	1	3	3	3	2	3	2	5	4
	Falex (scuffing load)	1–2	1–2	3	1–4	3	3	3–4	1–2	5	5
	Almen-Wieland (seizure load)	3	3	2	3	2	1	1–2	2	5	5
	FZG (A/16.6/90)	2	3	3	4	2	1	1	1	5	5
Wear behaviour	FZG (A/8.3/90)	2	1	1	1	1	1	1	1	3	4
	Reichert	2	2	2	3	3	3	2	3	5	5
	Timken; wear at 30 lbs.	4	4	4	2	5	1	2	1–2	5	5
	IfE* machine	5	5	4	2	1	2	1	1	2	4
	Falex: wear at 1000 lbs.	1	2	2–4	2	4	3	4	1	5	5
	Almen-Wieland; wear at 750 daN	4	4	1	2–3	2	1–2	2	3	5	5

* Machine of the Institute for Petroleum Research, Hannover.

Owing to the poor correlation between the operating parameters of laboratory devices and operating mechanisms it is not possible to use directly laboratory test results for service applications. It is obvious that the complex operating conditions of a hypoid gear cannot be completely reproduced by any laboratory method or device in present use.

Considering all the advantages and disadvantages of laboratory test methods it may be concluded that such devices are still essential for the initial development of new type lubricants. As the mechanism of friction under mixed or boundary lubrication conditions becomes more thoroughly investigated and understood the laboratory stage of assessment will become more important and complete mechanisms will only be used to finally check the results.

8.2.1 Improvement of the effectiveness of laboratory test devices

The improved accuracy of the results obtained with laboratory test devices and a better understanding of their operating conditions are closely related to progress in knowledge of the mechanisms of friction under mixed and boundary lubrication conditions, and of their dominant controlling factors. However, even on the basis of available knowledge in this field effectiveness of test devices can be improved by a thorough analysis of the operating conditions in the area of frictional contact and of the relationship between these conditions and the external parameters such as load, friction force, environment and lubricant temperature [95]. Several of these parameters will now be analysed.

Loading of the friction pair may be achieved by mass, by means of a spring or hydraulic pressure (Figure 8.8 *a*, *b* and *c*). Considered statically, all these systems are equivalent but because of the rotational motion of the moving part, dynamic effects should also be considered. Thus the loading system should be considered as an oscillating system of mass *m* and elasticity *C* (Figure 8.8 *d*). The moving part of the friction pair serves as a source of excitation having

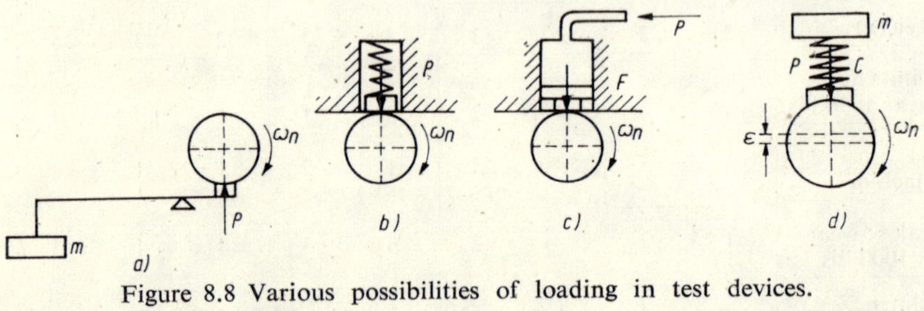

Figure 8.8 Various possibilities of loading in test devices.

a frequency ω_n and an amplitude, under rotational motion, equal to the eccentricity ε. If the load variation due to dynamic effects is sufficiently large, the friction pair may scuff although static values of the same quantity fall below the scuffing limit. Thus, two design characteristics of the device: the eccentricity and the dynamic characteristics, may affect indirectly the development of the friction process.

Another important aspect concerning the design characteristics of the loading system is the relationship between load and the pressure distribution over the contact surface. Only with a rigid loading system completely free of clearance is the pressure uniformly distributed over the contact surface. In real systems, with clearance, the pressure distribution may be influenced by deformation of the loading system (Figures 8.9 *a* and *b*). Although in both cases the total load is the same, in real systems, areas will exist where the contact pressure will be higher, which may affect the results, considering that wear rate and friction force usually do not depend linearly on pressure. The effect

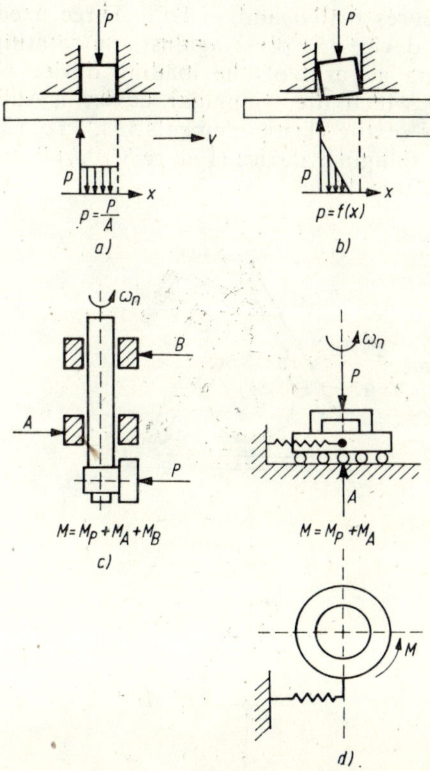

Figure 8.9 Effect of clearance on the pressure distribution in the contact region (*a* and *b*); methods of measuring the friction force (*c* and *d*).

of load concentration as a consequence of the clearance and elasticity of the loading systems may lead to appreciable differences in the results obtained for equal total loads with different devices. From this point of view it is advantageous to use point or line contact pairs instead of surface contact pairs.

The friction force and the coefficient of friction may not be measured or is measured with insufficient accuracy. Exact measurement of the friction force involves several difficulties as it acts only in the presence of a much larger normal loading force. In principle, the friction force may be measured in two ways: either by measuring the moment acting on the driving shaft (Figure 8.9 *c*) or by measuring the force acting on the fixed part by means of an elastic element (Figure 8.9 *d*). In the latter case, as the fixed part should transmit the normal force whilst moving freely in the direction of the friction force, the fixed part is mounted on a low friction device such as a ball bearing. In both cases, the measured moment differs from the friction moment M_p by the extraneous friction moments M_A and M_B of the bearings. The error introduced is not constant, its absolute value increasing with load.

To eliminate extraneous moments, a solution based on a new principle should be adopted (Figures 8.10 *a* and *b*) [90]. Three fixed parts (*2*) mounted in a triangular loading device (*3*) press against the rotating bush (*1*). Loading is achieved by displacing an arm of the loading device with a screw *P*. The moment of rotation to which the triangular device is subjected is controlled and measured by means of the elastic elements (*4*). From the diagram plotting the forces acting on the triangular device (Figure 8.10 *b*) it appears that only the friction moment acts on this device.

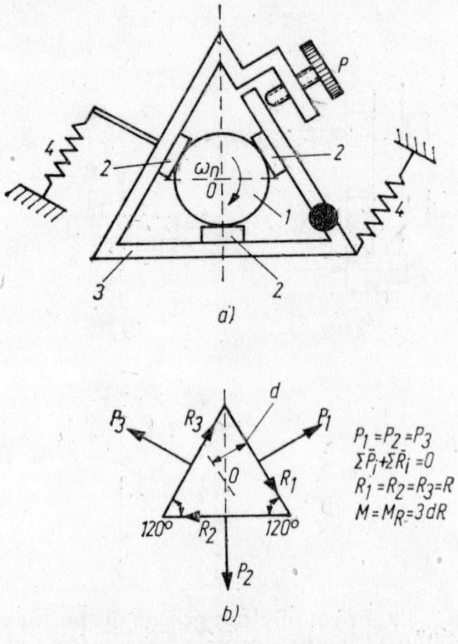

Figure 8.10 Triangular loading device used to eliminate parasite moments in the measurements of the friction force: *a*, schematic diagram; *b*, force diagram.

The environment in which frictional contact takes place, namely the lubricant and the gaseous atmosphere, are of primary importance in basic research on friction under conditions of mixed lubrication. Under such conditions, interaction between the lubricant and the atmosphere on the one hand, and the metallic surface on the other hand, is important. In addition to the use of a pure, chemically well-defined lubricant, it is necessary to ensure absolute cleanliness of the working area between test runs. This is difficult to achieve as the cleaning process can only be described qualitatively. The importance of lubricating fluid purity and the cleaning technique follow from the fact that some classes of sub-

9 Modern physical techniques in tribology

9.1 X-ray spectral analysis

9.1.1 Interaction of electromagnetic radiations with materials

When electromagnetic radiations penetrate a material several interactions occur between the radiation photons and the atoms and electrons of the material respectively, a process which causes attenuation of the beam. In the low energy range in X-ray spectral analysis or in diffraction phenomenon, the following interaction processes are important: the Thomson scattering, the Compton effect and the photoelectric effect.

In the Thomson interaction process the excited electron oscillates with the frequency of the incident radiation and dissipates the energy received in the form of an electromagnetic radiation of the same frequency as the initial radiation. With this type of interaction, the interacting electron does not receive kinetic energy. The Thomson effect makes diffraction phenomenon possible.

The Compton effect, which is a source of non-coherent radiations, takes place when a photon meets a free or slightly bound electron. The photon gives

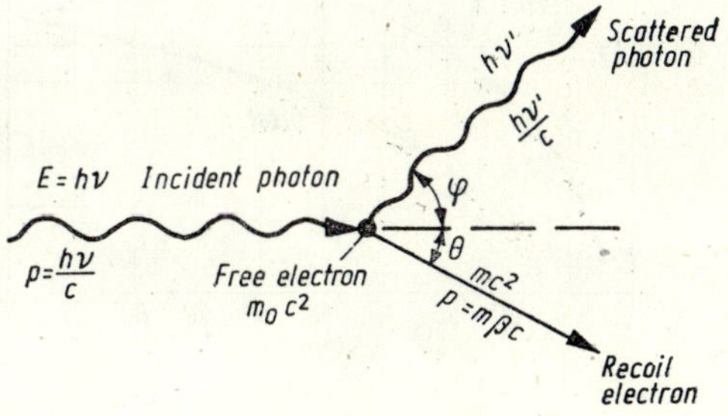

Figure 9.1 Reduction in the energy of a photon by the Compton effect.

up only a part of its energy to the electron, the former changing its direction and energy (Figure 9.1).

On the basis of the conservation of energy and momentum in interaction a relationship may be deduced between the increase in the wavelength $\Delta\lambda$ and the deviation agle φ

$$\Delta\lambda = \frac{h}{m_0 c}(1 - \cos\varphi), \tag{9.1}$$

or

$$\Delta\lambda = 0.024(1 - \cos\varphi), \tag{9.2}$$

if the values of Planck's constant (h), the rest mass of the electron (m_0) and the velocity of light (c) are replaced in relation (9.1). The Compton scattered radiation represents a fraction as much less as the incident radiation wave-length is greater or the angle which measures the scattered radiation is less.

The photoelectric effect is the process by which the photon expels an orbital electron giving up its entire energy to it. If the photon kinetic energy is $h\nu$ and the electron binding energy is Φ, then the kinetic energy T of the emergent electron is

$$T = h\nu - \Phi. \tag{9.3}$$

The process may take place only if $h\nu > \Phi$. The binding energies, functions of the atomic number for the orbital electrons K and L, most important in X-ray spectral analysis, are given in Figure 9.2 in the form of a continuous curve.

The probability of interaction by photoelectric effect increases as the photon energy is closer to the emergent electron binding energy. For this reason, a sudden increase in the photoelectric effect occurs for close energies. If the incident photons have a higher energy than that of the electrons K, the nearest to the nucleus, the photoelectric effect is predominant for this shell.

An atom brought to a higher energy state by the removal of an electron from its inner orbit is unstable; it will trap an electron from a neighbouring shell and will emit a photon the energy of which, characteristic of each atom, is determined by the energy difference between the two shells (Figure 9.3). It follows that the excitation energy should be higher than that of the emitted photon.

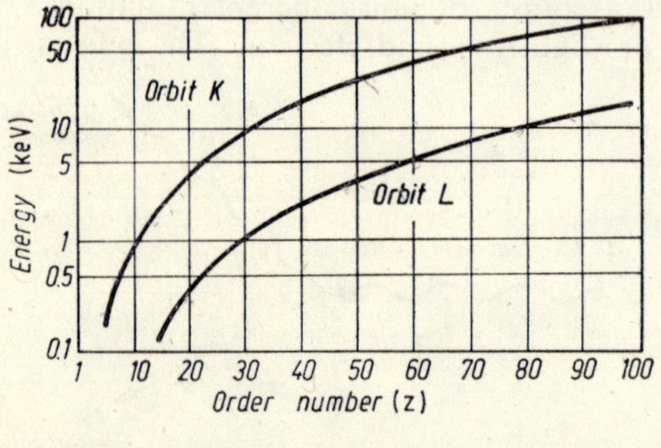

Figure 9.2 Binding energy of orbit electrons K and L.

As a consequence of the characteristic emission, the energy of the incident photon will be divided into a secondary diffused energy termed fluorescence radiation and an energy adsorbed in the form of photoelectric electron kinetic energy. From a quantitative point of view, the photoelectric effect is predominant in energies which are of primary interest for X-ray spectral analysis (Figure 9.4).

The X-characteristic spectral line may also be obtained by excitation of the atoms by means of an electron beam. Simultaneously with the characteristic radiation of the irradiated element a continuous spectrum is generated as a result of the deceleration of the electrons in the nucleus field (Figure 9.5). The energy distribution in the continuous spectrum depends on the maximum energy of the electrons in the incident beam and the target atomic number. The maximum energy of the continuous spectrum depends on the electron beam accelerating voltage by the relation

$$h\nu_{max} = eV, \tag{9.4}$$

where

V is the potential difference between the anode and the cathode,
e — electron charge,
h — Planck's constant.

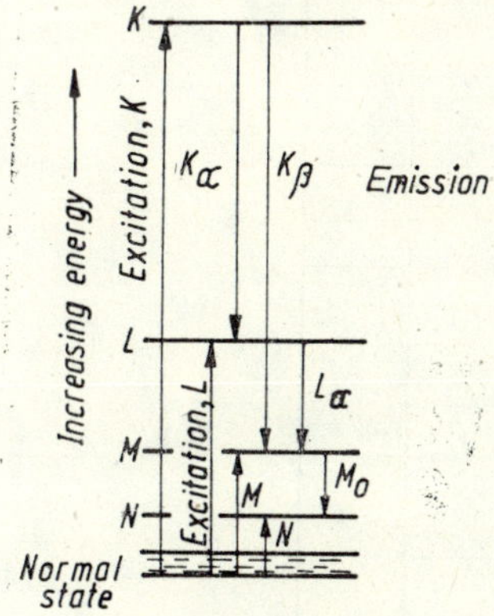

Figure 9.3 Energy levels in an atom and formation of characteristic radiation.

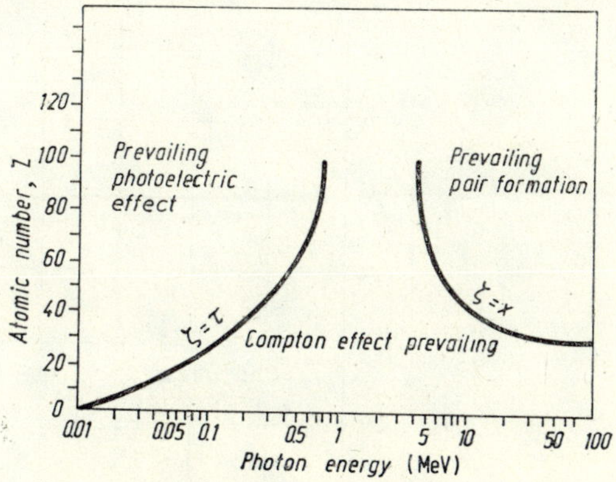

Figure 9.4 Types of electromagnetic radiation interaction with the substance in terms of energy and atomic number [49]. τ, ζ and $\varkappa$ respectively are the photoelectric, Compton and pair formation attenuation coefficients.

Of the characteristic radiations the K series is the simplest in structure. Generally, this consists of four lines of different intensities, denoted by α_1, α_2, β_1 and β_2. The relative intensity of the series K, L, M, etc., as well as of the lines in each series does not depend on the electron energy and the accelerating voltage, respectively. For the K series, the following ratio is found between the intensities of the component lines:

— light elements $\alpha_1 : \alpha_2 : \beta_1 : \beta_2 \approx 100 : 50 : 15 : 0.1$;
— medium elements $\alpha_1 : \alpha_2 : \beta_1 : \beta_2 \approx 100 : 50 : 28 : 6$;
— heavy elements $\alpha_1 : \alpha_2 : \beta_1 : \beta_2 \approx 100 : 54 : 26 : 9$.

9.1.2 Attenuation of a radiation beam

Due to interaction between the radiation and the material attenuation of the beam takes place. As absorption of a photon from the incident beam takes place in a single act and the number of interactions is always proportional to the number of remaining photons, the attenuation of the beam is a function of the distance travelled, and is of an exponential form

$$I = I_0 \, e^{-\mu x}, \tag{9.5}$$

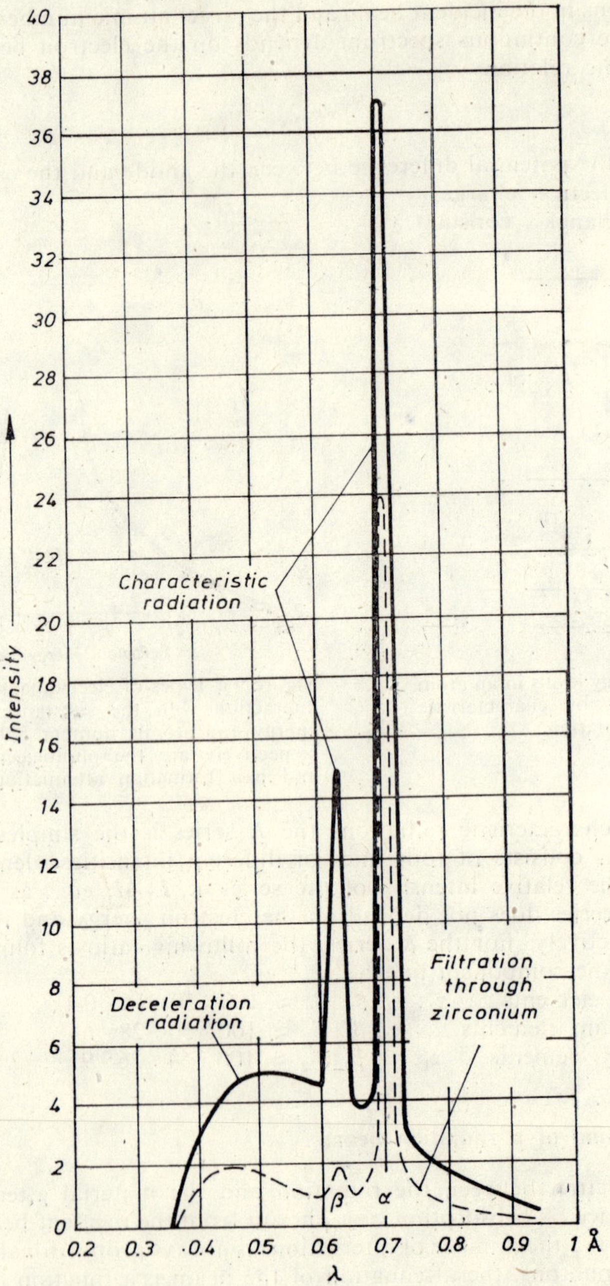

Figure 9.5 X-radiation spectrum for molybdenum target upon electron excitation [12].

where

I_0, I is the intensity of the monochromatic incident beam and at the distance x, respectively,

x — distance travelled,

μ — linear attenuation coefficient.

The absorption of a photon from the incident beam may, however, take place either by absorption or by scattering. Corresponding to these two processes, the coefficient of attenuation is the sum of two terms

$$\mu = \tau + \sigma, \tag{9.6}$$

where τ is the coefficient of absorption and σ is the coefficient of scattering. Both τ and σ represent the sum of the coefficients corresponding to the three modes of interaction: photoelectric effect, Compton effect and coherent scattering (at high energies other effects also occur, such as the formation of pairs, etc.).

If instead of the linear coefficient of attenuation μ, the mass coefficient of attenuation μ/ρ is used, where ρ is the substance density, the effect of the state of aggregation of the environment is eliminated. The mass coefficient depends only on the radiation wavelength and the absorbent atomic number.

The variation of the mass coefficient of absorption with the incident radiation wavelength λ and the atomic number Z of the absorbent has the form

$$\frac{\tau}{\rho} = c\lambda^3 Z^3. \tag{9.7}$$

This relation holds for $\lambda < \lambda_K$. If $\lambda \simeq \lambda_K$, a sharp decrease in the coefficient of absorption takes place as a consequence of the generation of the characteristic radiation. In the range $\lambda_K < \lambda < \lambda_L$ relation (9.7) holds again with another constant c. For $\lambda \simeq \lambda_L$ a new discontinuity occurs in the attenuation coefficient curve.

The mass coefficient of scattering σ/ρ depends on λ or Z only to a small extent and is, except for light atoms ($Z < 20$), much smaller than the coefficient

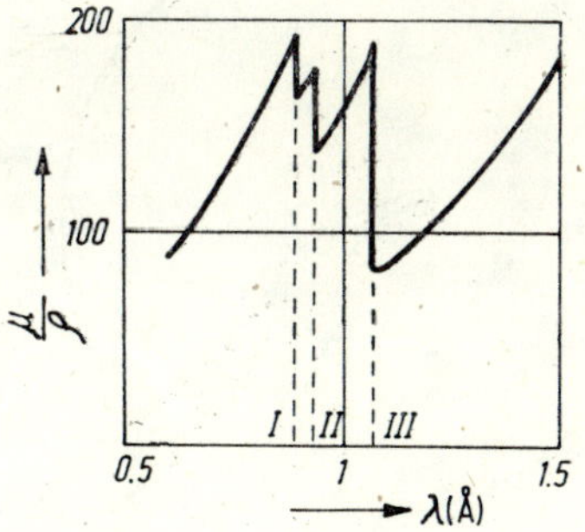

Figure 9.6 Mass attenuation coefficient of Pt versus wavelength.

of adsorption. Thus the coefficient of attenuation between discontinuities has the form

$$\frac{\mu}{\rho} = c\lambda^3 Z^3. \tag{9.7'}$$

Over the large range of energies of the incident photons the coefficient of attenuation has a characteristic form in which the cubic dependence is interrupted by the attenuation peaks (Figure 9.6).

9.1.3 Qualitative and quantitative analysis. Fluorescence analysis

Qualitative elemental analysis by characteristic X-radiation is based on two properties of the characteristic spectrum: simplicity (with respect to the optical spectrum) and the independence of the characteristic lines of the chemical bonds of the elements examined.

To generate the characteristic spectrum both the excitation of the atoms with electrons and X-rays are used. As a technique of elemental analysis the latter is preferred as the characteristic X-radiation is not accompanied by the continuous spectrum. Generation of the characteristic spectrum by means of an electron beam has been particularly developed for the technique of electron probe micro-analysis.

In principle, the fluorescence analysis technique uses an X-ray source to irradiate the specimen to be examined and a spectrometer, usually coupled with a radiation transducer (Figure 9.7). The spectral analysis of the characteristic spectrum is achieved by a rotation method in which the transducer is rotated at double speed with respect to the crystal speed. Thus, conditions are achieved so that all wavelengths are captured by the radiation transducer.

Due to the simplicity of the X-ray characteristic spectrum identification of the elements is relatively simple. As with optical spectroscopy, difficulties are encountered by the simultaneous presence of two elements having close spectra, one of which is in low concentration. For this reason, an element is considered to be identified only if at least two of its characteristic lines are identified.

The limit of detection depends on the specific conditions of examination. Generally, it is considered to be between 0.05 and 0.1% by weight. Under favourable conditions, it may be below 0.01%.

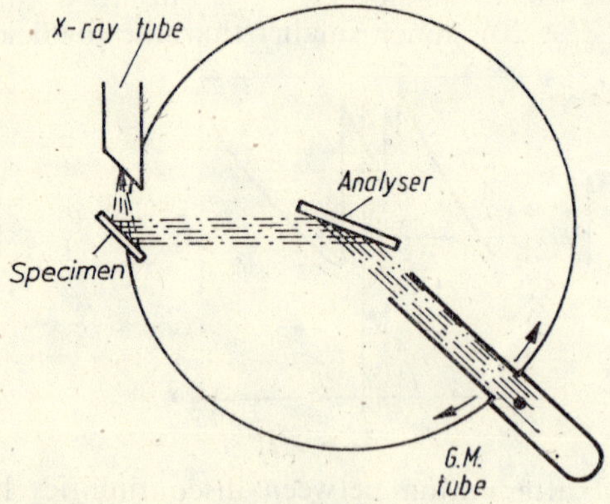

Figure 9.7 X-ray fluorescence analysis.

By measuring the intensity of various spectral lines data may also be obtained on the concentration of the elements examined. It must be borne in mind that, as distinct from optical spectroscopy, the intensities of the characteristic lines do not depend simply (linearly or logarithmically, for instance) on the element concentration. This complication is attributed to the fact that the intensity of the characteristic lines depends to a large extent on the primary electron energy and the primary radiation wavelength respectively and also on the selective adsorption of the primary and characteristic radiations in the specimen analysed.

The intensity J_f of a fluorescence characteristic line of wavelength λ_f depends on the intensity and wavelength of the primary radiation (J_0, λ_p) and the number of atoms participating in the emission N_f

$$J_f = \frac{J_0(\lambda_p)\, Q_f(\lambda_p) N_f A}{\dfrac{\mu(\lambda_p)}{\sin \varphi} + \dfrac{\mu(\lambda_f)}{\sin \psi}}, \tag{9.8}$$

where

$Q_f(\lambda_p)$ is the probability of emission,

A — specimen surface,

φ and ψ — incidence and emission angle, respectively.

If the specimen is sufficiently thick so that the primary radiation is completely adsorbed, then the intensity of the fluorescence line is a function of the ratio λ_p/λ_f.

The usual technique for quantitative analysis is to measure several standards of known specific composition to provide standard calibration curves of concentration versus intensity of a spectral line. Account should be taken however, that such calibration is valid only for the composition of the standard specimen.

Another method is to add to the analysed specimen a known quantity of an element having an atomic number close to that of the element to be examined. By measuring the intensities of the main lines in both elements the concentration ratio can be determined directly. The difference in the efficiency of the transducer for the two lines measured should also be taken into account.

Despite the difficulties encountered, X-ray fluorescence analysis has been progressively developed. Vacuum operating facilities are available which permit elementary analysis to be performed up to $Z = 11$. The size of the specimen examined can be reduced to the order of a few milligrams by using a spectrometer with a curved crystal. The simultaneous use of several transducers enables the main elements of a specimen to be determined concomitantly.

X-ray Photoelectron Spectroscopy. The energy of the electrons resulting from the photoelectric effect depends on electron binding energy and on the energy of the incident X-radiation (relation (9.3)). By measuring the energy of the photoelectrons by means of a spectrometer, the energy of incident X-radiation being known, the binding energy of the photoelectrons can be determined. The binding energy of the orbital electrons is characteristic of each atom. However, this is affected by the presence of other atoms in the molecule. This measurable change in the binding energy of the electrons permits identification in certain instances both of the atoms and of the compounds in which these atoms are included. This advantage, together with a low penetration power (1 ÷ 2 nm) makes the use of this method attractive for the analysis of the thin layers on frictional surfaces [201, 202].

Applications. For the study of worn surfaces or wear particles qualitative and quantitative X-ray fluorescent analysis has several advantages:

— the non-destructive character of the method is important, especially in the study of surface layers,

— a relatively small volume of material only is required for analysis which is important in the analysis of surface layers and wear debris,

— it is a relatively simple procedure.

Although the method can provide only a quantitative elemental analysis, it may also be used for the study of the surface layer structure. For this it is necessary to determine the percentage composition of the additive used and the layer composition. Differences in the percentage composition provide indications on the mechanism of formation of the layer: by adsorption or by reaction.

This method was used by Godfrey, for example, for the study of the mechanism of action of several additives [71]. Details concerning the method itself can be found in references [133, 62].

9.2 X-ray diffraction

9.2.1 Crystalline structure of solids

The normal state of a solid is the crystalline state. In a crystal the atoms occupy well-determined, fixed positions in relation to each other. The ordinate microscopic structure will also be reflected in the macroscopic structure of the crystal conferring to it a certain symmetry. The macroscopic symmetry manifests itself by the existence of three elements of symmetry: symmetry axes, centres and planes.

Most solid substances are polycrystalline consisting of a compact agglom. eration of crystals of small dimensions having nearly perfect crystallinity- Due to the random orientation of the crystallites these materials are isotropic. Cases exist in which the crystallites have a preferred orientation as a consequence of a processing procedure.

Unit cell. The distribution of the atoms in a crystal can be described by means of a three-dimensional system of coordinates. By dividing the space into a lattice of straight, equidistant lines which are parallel to the three axes, a lattice is obtained which has the unit cell as an irreducible element. There are only 14 types of lattices by repetition of which all crystal structures can be obtained (Bravais lattice). The basic condition of a Bravais lattice is that each point of the lattice must be arranged so that the surroundings of any point is the same as any other in the lattice.

To describe unit cells 7 types of lattices are required, possessing a minimum of symmetry elements. These lattices coincide with the macroscopic crystal systems (Table 9.1 and Figure 9.8).

If a lattice has only one lattice point per unit cell, it is said to be a "primitive lattice". It is possible to describe the fourteen Bravais lattices entirely in terms of primitive lattices. However, it is much more convenient to describe them in terms of "non-primitive" lattices, i.e. lattices with more than one point

Table 9.1

Characteristics of crystal systems

System	Basic symmetry	Unit cell
1	2	3
Triclinic	None	$a \neq b \neq c$ $\alpha \neq \beta \neq \gamma \neq 90°$
Monoclinic	One 2-fold rotation axis	$a \neq b \neq c$ $\alpha = \gamma = 90°; \beta \neq 90°$
Orthorhombic	Three 2-fold rotation axes	$a \neq b \neq c$ $\alpha = \beta = \gamma = 90°$
Rhombohedral	One 3-fold rotation axis	$a = b = c$ $\alpha = \beta = \gamma \neq 90°$
Tetragonal	One 4-fold rotation axis	$a = b \neq c$ $\alpha = \beta = \gamma = 90°$
Hexagonal	One 6-fold rotation axis	$a = b \neq c$ $\alpha = \beta = 90°; \gamma = 120°$
Cubic	Four 3-fold rotation axes Three 2-fold rotation axes	$a = b = c$ $\alpha = \beta = \gamma = 90°$

per unit cell. Table 9.2 describes the fourteen Bravais cells on the basis of seven crystal systems. It is noted that, in addition to the primitive cells, cells with a lattice point in the centre of the cell, cells with lattice points in the basic faces or cells with lattice points in all faces also occur.

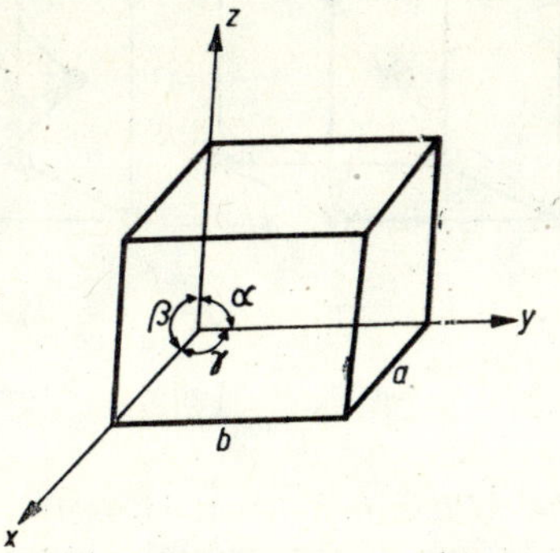

Figure 9.8 Triclinic lattice unit cell.

The most frequently encountered metal lattices are the face-centred cubic, the body centred cubic and the hexagonal lattice (Figure 9.9 and Figure 9.10). It should be noted that the hexagonal unit cell is a parallelepiped having sides a, b and c and an angle $\gamma = 120°$. The hexagonal prism is obtained by joining

Table 9.2

Bravais cells

System	Bravais lattice	D'Herman-Mauguin symbol
Triclinic	simple	P
Monoclinic	simple base-centred	P C
Orthorhombic	simple base-centred face-centred centred	P C F I
Rhombohedral	simple	R
Tetragonal	simple centred	P I
Hexagonal	simple	P
Cubic	simple face-centred centred	P F I

two elementary full prisms and two semi-prisms. Certain metals have a compact hexagonal structure which, in addition to the atoms from the prism vertices, also contains an atom at the height $c/2$ in the centre of the semi-prism.

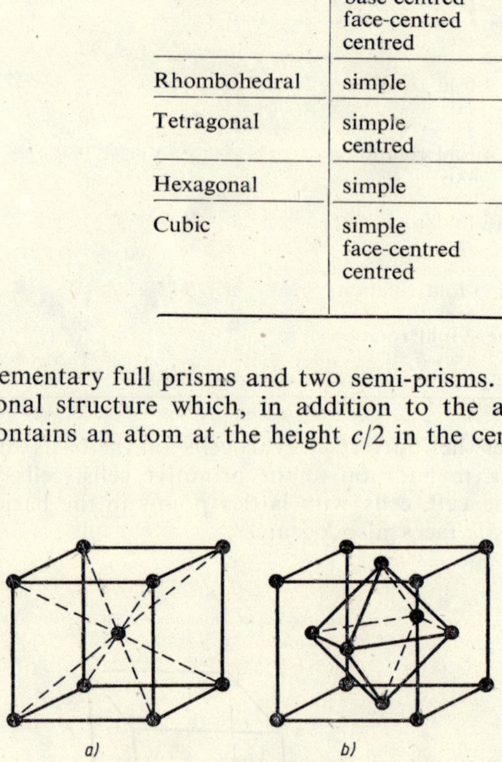

Figure 9.9 Unit cell:

a, centred cubic lattice; *b*, face-centred cubic lattice.

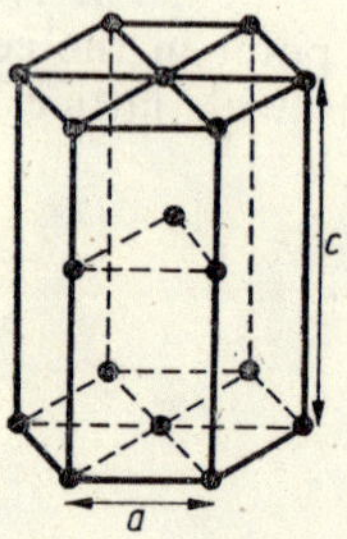

Figure 9.10 Unit cell. Compact hexagonal lattice.

The Miller indices. It is useful for the study of diffraction phenomenon to use a notation of the crystal planes which only defines their orientation with respect to the coordinate axes without considering their space position. From this, a parallel set of crystal planes, one of which passes through the origin and the others through the coordinate points a/h, b/k, c/l where h, k and l are integers, is denoted by $(h\,k\,l)$. The integers h, k, l are called the Miller indices. The notation for the plane (111) is shown in Figure 9.11.

Interplanar spacings. With the aid of Miller indices interplanar spacings can be defined for all crystal systems. In the cubic system the interplanar spacing of the planes $(h\,k\,l)$ is

$$d_{(h\,k\,l)} = \frac{a}{\sqrt{h^2 + k^2 + l^2}}, \tag{9.9}$$

while in the hexagonal system

$$\frac{1}{d^2_{(h\,k\,l)}} = \frac{4}{3}\,\frac{h^2 + hk + k^2}{a^2} + \frac{l^2}{c^2}. \tag{9.10}$$

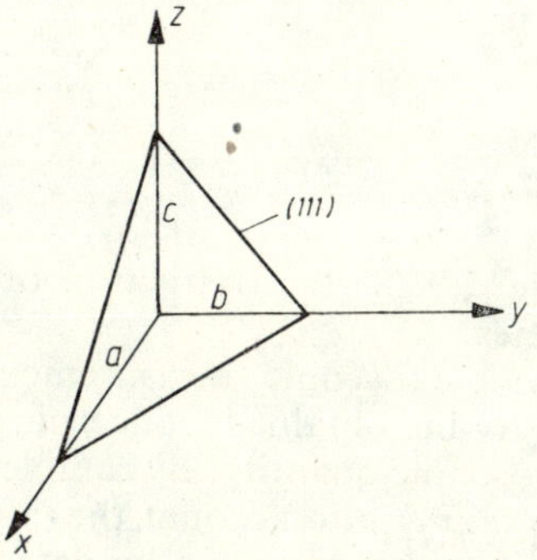

Figure 9.11 Miller indices for the (111) plane.

The interplanar spacing represents the geometrical characteristic by means of which a crystal lattice participates in the process of diffraction.

The reciprocal lattice of a crystal is a lattice of points, each point representing a whole set of parallel atom planes within the crystal. Each reciprocal lattice point is situated on a line perpendicular to the corresponding planes of the real crystal, and at a distance from an arbitrarily chosen origin, equal to the reciprocal of the spacing between the planes (Figure 9.12). The vector of the reciprocal lattice is given in magnitude by the relation

$$d^*_{(h\,k\,l)} = \frac{K}{d_{(h\,k\,l)}}, \tag{9.11}$$

where K is arbitrarily taken to be λ for interpreting X-ray diffraction patterns, and unity for interpreting electron diffraction patterns. Construction of the reciprocal lattice greatly facilitates the interpretation of diffraction patterns.

9.2.2 Bragg's law. Diffraction techniques

By the interaction of an electromagnetic radiation beam with the material, secondary radiations occur, some of which have the same frequency as that of the incident beam (coherent diffusion). As with visible radiation, coherent radiation makes possible the occurrence of the interference phenomenon.

If a plane, monochromatic X wave meets a crystal lattice, then the condition that the diffracted beam is intensified is given by Bragg's law (Figure 9.13)

$$2d_{(h\,k\,l)} \sin \theta_{h\,k\,l} = \lambda, \tag{9.12}$$

when $AB + AC = n\lambda$.

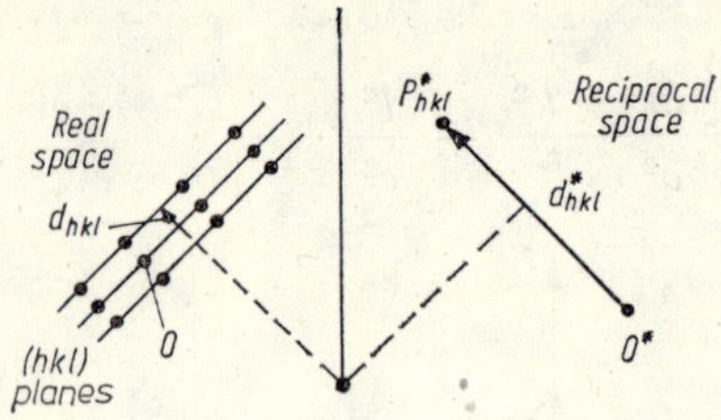

Figure 9.12 Reciprocal lattice definition.

In relation (9.12) the diffraction order is included in the Miller indices of the set of planes.

From a macroscopic measurement of the angle θ, the Bragg relation permits data to be obtained concerning the characteristics of crystal lattices ($d_{h\,k\,l}$), a microscopic quantity. By completely interpreting the diffraction patterns, that is by also taking into account the intensities of the diffraction lines, all parameters of the crystal lattices can be determined.

In most applications, the materials examined are polycrystalline, their small crystals being in a random or preferred orientation. For a random orientation

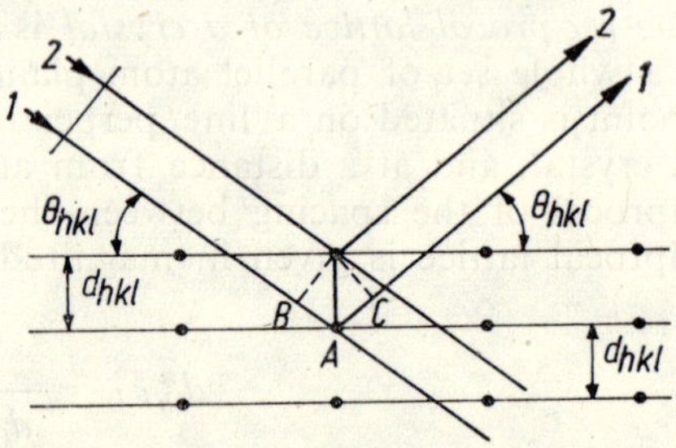

Figure 9.13 Diffraction of a coherent beam on a set of planes.

the diffraction pattern exhibits a circular symmetry with respect to the incident beam. A preferred orientation is reflected in the distribution of the diffracted beam intensity around the circumference.

The X-ray diffraction techniques most used in tribological applications are: the flat-plate X-ray diffraction camera and the cylindrical powder camera techniques. In both cases a monochromatic X-ray is used.

The first technique has three variants, forward transmission, back reflection and reflection at a glancing angle. This last technique has the advantage of permitting direct examination of a worn surface (Figure 9.14). A collimated monochromatic X-ray beam is made to impinge on the worn surface at a small angle α. Thus, the specimen body obstructs the direct access of the primary beam to the photographic plate. The resulting diffraction cones from the solid polycrystalline surface determine diffraction rings by the intersection with the plane of the photographic plate. With this technique as all the surface participates in the formation of the diffraction pattern, the rings obtained on the photographic plate are somewhat diffused. This effect can be overcome by setting the

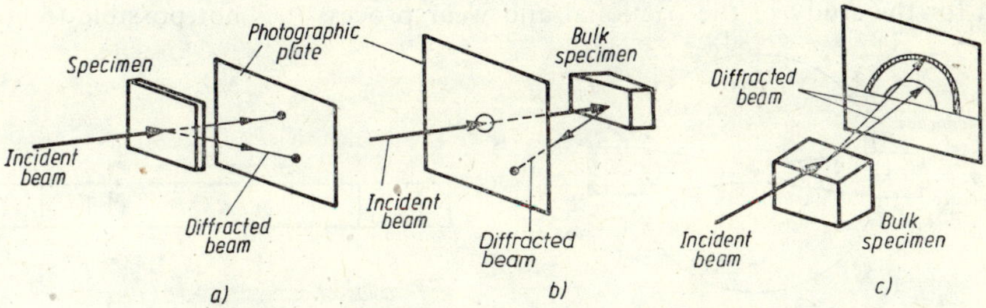

Figure 9.14 Techniques for obtaining interference patterns: *a*, transmission; *b*, reflection back; *c*, reflection at glancing incidence.

specimen so that the X-ray beam impinges at an edge instead of the centre of the flat specimen (the edge is, however, not always significant of the frictional process).

With the powder camera technique the specimen to be analysed is contained in a capillary tube placed in the centre of a circular camera (Figure 9.15). By one of the camera diameters an X-ray beam is introduced near to the specimen by means of a protecting tube. The emergent beam is taken out of the camera also by means of a protecting tube. This arrangement aims at protecting the film against direct radiation effects. A photographic film is placed at the lateral inner side of the circular camera.

The diffraction cones of various vertex angles (2θ) will produce on the film several curved lines having their centre of curvature at the intersection of the incident beam with the surface of this camera (Figure 9.16). The camera radius being known, the angle 2θ can be calculated directly for each diffraction cone, by measuring the position of the diffraction lines with respect to the intersection point of the radiation beam with the film.

The principal use of the powder camera is for the identification of compounds in a mixture. The angles 2θ are determined, then the interplanar spacings $d_{(h\,k\,l)}$ are calculated by the Bragg relation (the wavelength of the radiation used being known). These values, together with the corresponding intensities of the diffraction lines measured on the film permit, with the aid of some similar quantities determined by calibration, establishment of the composition of the specimen analysed.

Determination of the composition of a mixture is difficult owing to:

(a) the existence of multiple diffraction even for a single substance,

(b) the limited resolution of the diffraction camera,

(c) deficiencies in the method of measurement of the diffraction line intensity on the film.

Considering, however, that in most cases, completely unknown substances are not being examined, this technique may offer valuable data concerning composition.

Applications. The most frequent tribological application of X-ray diffraction is for the analysis of wear debris. The main reason for its use is the relatively small amount of substance required for analysis as from most laboratory devices used for the study of the frictional and wear process it is not possible to collect

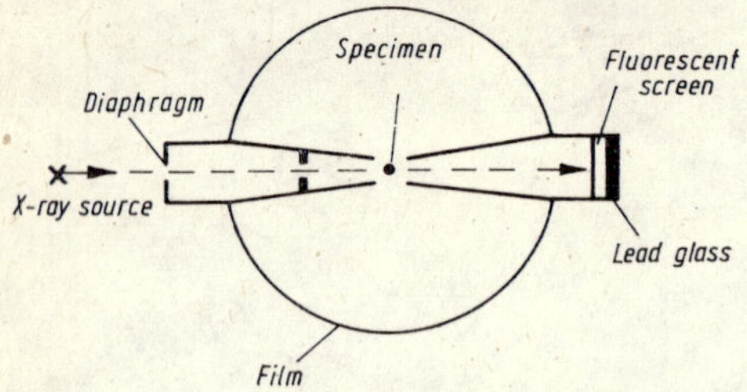

Figure 9.15 Diffraction powder camera.

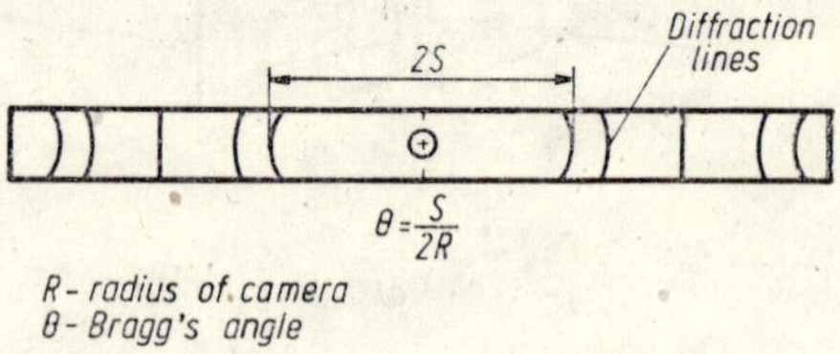

Figure 9.16 Diffraction lines obtained by powder camera techniques.

large quantities of wear debris. Also, the smaller the working surface of the pair, the smaller is the quantity of wear debris, a situation which occurs under line or point contact with concentrated load. Under conditions of lubricated frictional contact the wear rate is also decreased.

This technique was used, for instance, to examine the composition of wear debris produced by the dry frictional contact of stainless steel in air [144].

For the analysis of the surface layer formed on a friction surface the layer is first stripped from the support surface by an electrolytic technique. The composition of the layer formed on steel surfaces by the use of extreme pressure additives was determined in this way [71].

The technique can be also used to determine directly the composition without removal from the surface only if the layer is sufficiently thick. If the layer is not thick enough the relatively high penetrating power of X-rays (of the order of 10^{-2} mm for metallic surfaces) causes the diffraction pattern of the layer to be superposed over that of the support material.

By means of a special technique, Figure 4.17, X-ray diffraction patterns are possible from adsorbed (chemisorbed) thin layers. Thus, the preferred orientation of the fatty acid molecules on metallic surfaces was determined.

The high penetrating power of X-rays (as compared with that of electrons) permits, however, analysis of the effect of friction on the orientation of crystallites in the material surface layer. The anisotropy of the surface layer can be established by the flat-plate diffraction camera. Due to preferred orientation of the crystallites, the diffraction rings have interference maxima around the circumference. Generally, all metal processing techniques result in an orientation of the crystallites in the bulk of the material in the form of a fibrous or a lamellar texture. The frictional process forms a lamellar-type surface texture.

Use of this technique has shown that with a graphite on copper pair the copper surface layer also exhibits a specific orientation; the (321) planes of the copper crystals tend to become parallel to the frictional surface. It is possible that this preferred orientation of the copper surface influences the orientation of the graphite crystallites [143]. A similar phenomenon was also noted following dry frictional contact or abrasion of aluminium and steel [129]. Two types of layer were identified: a heavily deformed thin surface layer followed by a thicker layer in which the crystallites were only partly deformed. A change in the composition of steel affected only the second layer.

Another application of X-ray diffraction is the study of the wear of monocrystals. Determination of the crystallographic planes and crystal cutting is based on the measurement of the diffraction pattern. Investigation of the behaviour of monocrystals has revealed that the coefficient of friction is always lower in the plane of maximum atom density [159].

9.3 Electron microscopy

9.3.1 The transmission electron microscope

The most important characteristic of a microscope is its resolution. This quantity is defined as the minimum distance between two points for which the instrument still provides resolute pictures which, although superpose partly, can be distinguished as being definite at the sufficient magnification. The microscope magnification, in the case of the optical microscope, is given by the product of the objective lens magnification and the ocular magnification. The magnification power should, however, be adapted to the objective lens resolution power. Irrespective of the amount of ocular magnification it will be capable of resolving two neighbouring points only if these have previously been resolved by the objective lens in the intermediate image.

The resolution power d of an objective lens depends, however, only on the objective aperture 2α, the wavelength of the light used λ and the refractive index n of the medium between the object and the objective

$$d = \frac{\lambda}{n \sin \alpha} \cdot \tag{9.13}$$

The largest aperture possible is $2\alpha = 144°$, i.e. $\sin \alpha = 0.95$. The refractive index of air is 1 but may increase to 1.518 with cedar oil or to 1.734 with bromonaphthalene. Use of ultraviolet light instead of white light gives the maximum resolution of the optical microscope $d_{\min} \simeq 2.5 \cdot 10^{-4}$ mm. In normal light, the minimum resolving distance increases to about $5 \cdot 10^{-4}$ mm.

Another characteristic of the microscope is the depth of field, namely the distance in the direction of the microscope optical axis, between the extreme planes for which the image is still clear. The dependence of the depth on the numerical aperture of the objective and the magnification is approximately inversely proportional. For a magnification of 1000 × and a numerical aperture of $\sin \alpha = 0.95$ the depth of field is only 0.075 μm.

From relation (9.13) the only possible way to increase the resolution is to reduce the wavelength of the radiation used. Moreover, this radiation should also obey two conditions: it must be capable of being readily directed and it must interact with the substance. All these conditions are satisfactorily fulfilled by an electron beam.

A wave λ is associated with a monoenergetic electron beam characterized by the impulse p

$$\lambda = \frac{h}{p}, \tag{9.14}$$

where h is Planck's constant.

Expressing the impulse p in terms of the accelerating voltage V of the electron beam and using the numerical value of h the following relation is obtained

$$\lambda = \left(\frac{150}{V}\right)^{1/2} \cdot 10^{-7} \text{ mm.} \tag{9.15}$$

For an accelerating voltage of 100 keV, which is necessary to ensure a sufficient specimen penetration, a wavelength equal to $\lambda = 4 \cdot 10^{-9}$ mm is obtained as compared with that of visible light of $\lambda = 4 \cdot 10^{-4}$ mm. However, such a large improvement in resolution is not possible due mainly to spherical aberration of the magnetic lenses, which necessitates an important reduction in the numerical aperture of the objective. The magnetic lens has an aperture of only 10^{-2} rad. Consequently, the resolving power is reduced to about $4 \cdot 10^{-7}$ mm, i.e. still 1000 times greater than that of the optical microscope. Due to the small aperture angle the electron microscope has a much greater depth of field than the optical microscope.

In order to direct the beam, both magnetic and electrostatic lenses may be used. Due to certain favourable characteristics, electromagnetic deflection electron microscopes are largely used (Figure 9.17).

A sharp wolfram tip heated by a filament is used as an electron source. To reduce as much as possible the cathode surface, the whole filament is enclosed in a perforated cylinder having a negative potential with respect to the cathode. This assembly allows a concentration of the electron beam in a small area ahead of the filament, which thus becomes the virtual cathode of the source. From the virtual cathode the electrons are accelerated by means of a perforated anode, placed in the immediate vicinity. The construction of the anode-cathode system finally produces a narrow and homogeneous electron beam.

As with the optical microscope, a condenser lens concentrates the beam of electrons onto the specimen. Due to its structure, the specimen, which should be very thin, produces a modulation of the electron beam. Further, the objective lens produces a first image magnified about $100 \div 300$ times. Finally, a portion of the intermediate image is magnified by a projector lens on a fluorescent screen or a photographic plate. Ahead of each lens there are diaphragms limiting the solid angle of the beam. The whole trajectory of the electrons, from the filament to the fluorescent screen, and the specimen are in vacuum ($\sim 10^{-4}$ torr) to eliminate electron attenuation and scattering. Specimen illumination is varied by the diaphragm and the current through the condenser lens while the magnification is varied by the currents through the objective and projector lenses.

To produce a high quality electron microscope a very accurate construction is essential. For instance, the accuracy required for the mechanical construction of the microscope column is such that in a length of more than 1 m an alignment of the various subassemblies of several tens of microns should be possible. Another important requirement is stability of the electrical parameters, as the final image is obtained photographically. Any change in the position or size of the image during exposure reduces quality although the instantaneous performance of the instrument may be high.

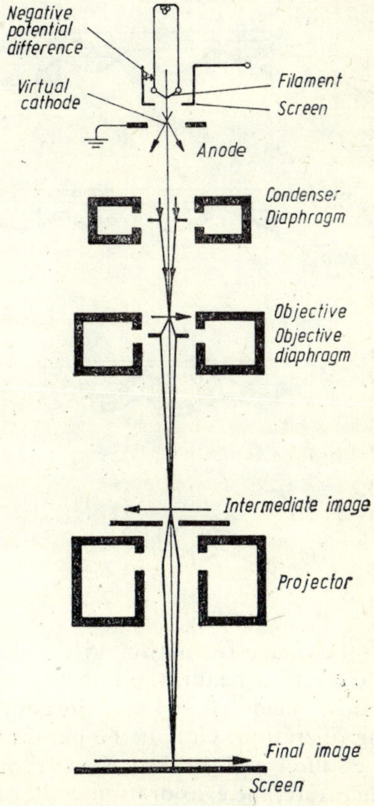

Figure 9.17 Schematic diagram of a transmission electron microscope with electromagnetic optics.

A characteristic of transmission electron microscopes which has caused problems in its application to the study of surfaces is the low penetrating power of the electron beam. To overcome this disadvantage surface replicas are used. The necessary introduction of this intermediate operation has several disadvantages:

(a) because of brittleness due to its small thickness the replica must be supported on a grid, which means that parts of this replica are masked by the opaque grid bars,

(b) the multiple operations involved in replica manufacture may alter its resolution,

(c) however careful the manufacture of a replica, the capacity of this film to reproduce surface details is limited.

Details in the replicas smaller than $5 \cdot 10^{-6}$ mm are certain to be non-representative of the real surface. The replica may be obtained either directly as a negative or in two stages. The two-stage positive replica is more often used to facilitate interpretation of the image.

To manufacture a replica the surface under examination is covered with a varnish layer, which, after it is dried, contains details of the surface on which it was formed. Then the film is carefully removed from the surface, so as not to damage the useful surface. To improve the contrast of the electron image the replica surface is shadowed, that is a layer of metal of high atomic number is deposited in a certain direction by vaporizing in a vacuum. A layer of carbon or other structureless material is evaporated in vacuum, normal to the surface. Finally, the plastic layer is dissolved away and the carbon replica is deposited on a microscope grid (Figure 9.18).

The method of making the replica should be taken into account in the interpretation of the resulting image. The formation of the electron image is based on the different attenuation of the electron beam in the replica cross section. As the basic material of the replica, for instance carbon, attenuates only slightly the electron beam, the shadowing material provides the main contribution

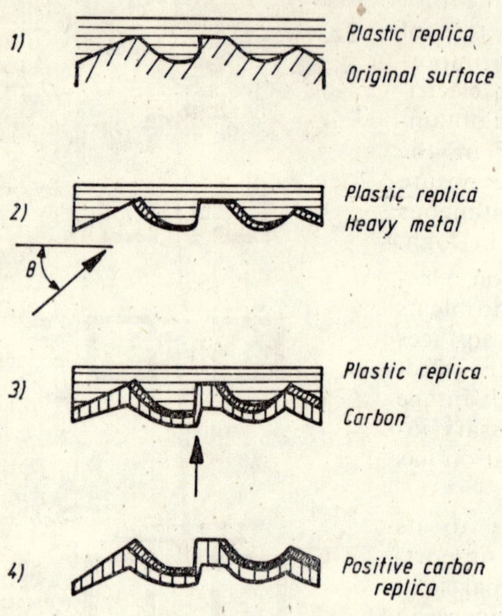

Figure 9.18 Stages in manufacturing a positive carbon replica.

to the image formation. Although the replica is placed in a direction normal to the electron beam, the image is formed by oblique illumination due to the angle used in shadowing. Thus, the contrast in the image depends largely on the shadowing operation. Only if the parameters of this operation (e.g. the distance between the source and the film, the evaporation rate, the pressure value in the evaporation enclosure, the evaporation angle) are kept constant, will the results be reproducible.

Another difficulty in the interpretation of the images obtained by this technique is that it is not possible to observe successively large and small surfaces on the same replica. This would permit location and correlation of the observed specific features. It is also difficult to locate a certain area of the real surface on the replica. The advantage of replicas is that it is a non-destructive technique which permits observation of the evolution of a surface during the various stages of the wear process.

Applications. Despite the above difficulties, transmission electron microscopy has been successfully applied to the study of surface changes during friction and wear processes. Any conclusions, however, should be correlated with the operating conditions as the microscopic examination of surfaces is only an auxiliary research method. Numerous applications include the study of changes in bearing lining in terms of lubricant composition, the study of metallic surfaces during frictional contact under conditions of boundary lubrication, the study

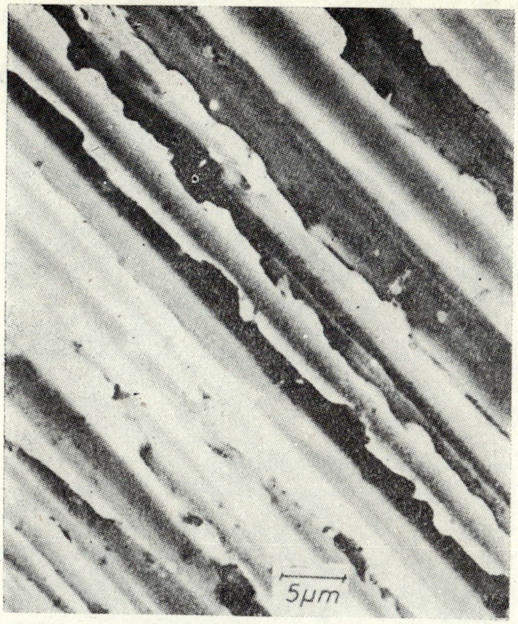

Figure 9.19 Transmission electron microscope picture. Surface of a fresh pin in an Almen-Wieland machine [2].

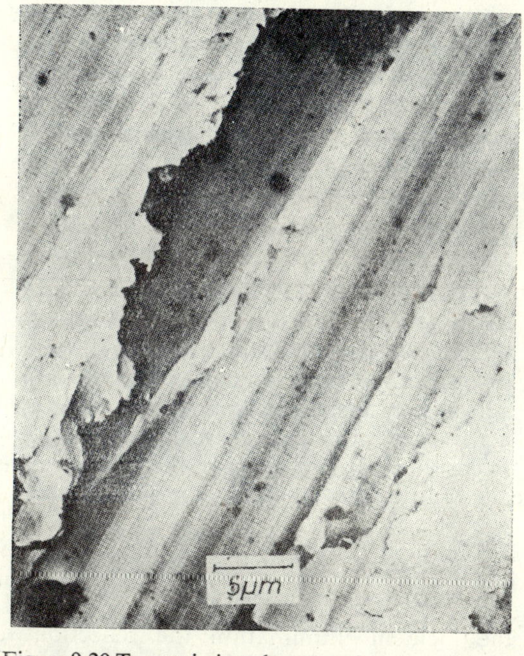

Figure 9.20 Transmission electron microscope picture. Surface of the pin shown in Figure 9.19 after running in paraffin oil under 400 daN[2].

of the mechanism of friction of polytetrafluoroethylene and the oxidative mechanism of friction of steel [177, 2, 160, 142]. Figures 9.19 and 9.20 show the pin surface of an Almen-Wieland pair before operation and after running in clean mineral oil.

9.3.2 **Reflection electron microscopy**

The intermediate stage of replica manufacture may be avoided by the use of a reflection technique when the surface under examination is placed directly in the microscope specimen chamber (Figure 9.21). The electron beam from the

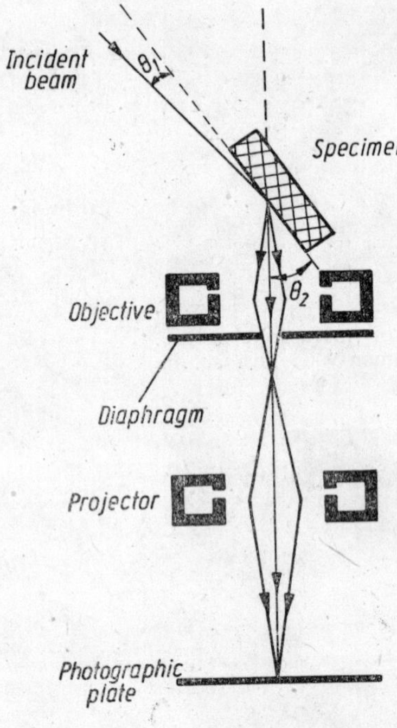

Figure 9.21 Schematic diagram of reflection electron microscope.

condenser lens illuminates the surface at an angle θ_1 to the surface, which, in turn, is viewed at an angle θ_2 (the angle of the microscope axis to the examined surface). To obtain maximum resolution, a certain relationship should exist between θ_1 and θ_2. In any case the resolving power cannot be better than $2.5 \cdot 10^{-5}$ mm. The usual values for θ_1 are 1° to 6° and for θ_2 6° to 23°. It should be noted that the image is less bright than in the case of transmission as owing to the small angle θ_2 only a small part of the intensity of the initial beam is utilized.

Generally, transmission E. M. can be used for reflection images. The tilt of the incident spot is achieved either by tilting the cathode and condenser lens, or by magnetic deviation of the beam.

Because of the incident illumination the reflection technique provides improved contrast for the examination of the surface microtopography. It should be noted, however, that, due to the mode of image formation, its quality is not uniform over the whole surface, especially at large magnifications. The reflection technique is now superceded by emission microscopy and scanning electron microscopy.

9.3.3 The emission electron microscope

The stimulated electron emission of a surface may be used to obtain an electron image. The specimen under examination is the cathode and the magnified image is produced on the fluorescent screen or photographic plate. Emission microscopes use either electrostatic or magnetic lenses.

To obtain electron emission from the surface under examination the thermionic effect, ion or electron excitation are used. With thermionic emission, the specimen is usually heated by an electron beam (on the opposite side of the surface to be examined) until free electrons are produced. The different emission capacity of points on the surface depends on the material and its crystal orientation, as well as on the presence of some layers. With specific substances the electron releasing energy may be reduced so that the surface examination can be commenced at a temperature of 700°C. The specimen may be heated to about 2000°C, if necessary, so that this technique is useful for examining structural changes in alloys, chemical surface processes and diffusion processes.

If it is essential to keep surface temperature at a low value, the electrons are generated by ion excitation. The ion beam is obtained separately and accelerated onto the specimen by a voltage of about 20–80 keV. An additional

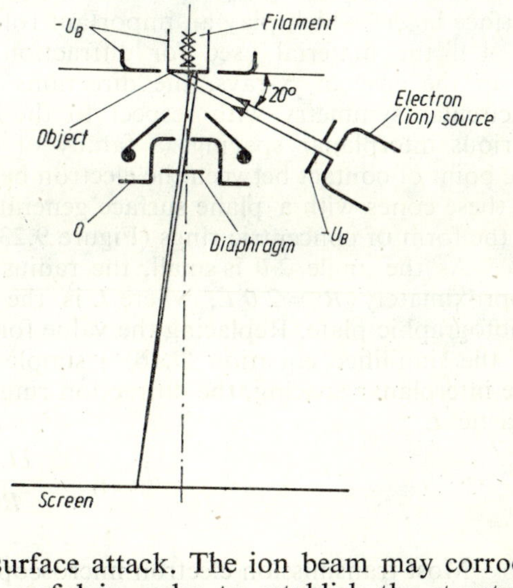

Figure 9.22 Schematic diagram of emission electron microscope.

possibility offered by this variant is surface attack. The ion beam may corrode the surface, which in some cases is useful in order to establish the structure of the material.

When electrons are used as the energy source, the image is formed by secondary electrons. Of the three techniques, the last affects the surface least.

Usually, emission microscopes are constructed to permit the use of any technique of obtaining electrons.

The resolving power is from 140 Å to 1400 Å, depending on the type of objective lens used. Figure 9.22 shows the schematic diagram of an emission microscope with an electrostatic objective lens in which electron emission is achieved by one of the methods described. Figure 5.11 shows the image of a wear scar on a steel surface, obtained with an emission electron microscope by activation with electrons.

9.4 Electron diffraction

A parallel monochromatic electron beam may be considered to be equivalent to a plane wave of wavelength λ given by relation (9.14). At the accelerating voltages used λ is usually much less than d, so the Bragg relation may be written in a simplified form

$$2d_{(h\,k\,l)}\theta \simeq \lambda. \tag{9.16}$$

An interplanar spacing $d = 1$ Å and an accelerating voltage of 100 keV gives an angle of incidence of $\theta \simeq 1°$.

Due to the small angle of incidence and the much stronger interaction of the electrons with the material, the depth of penetration of the beam in a direction normal to the surface is much smaller. Consequently, it is considered that the thickness of the layer which contributes to the formation of the diffraction pattern in the case of very smooth surfaces is of the order of several tens of Angströms. This characteristic makes electron diffraction very useful for the study of thin surface layers which play an important role in friction of solid surfaces.

If the material used for diffraction is crystalline and unoriented, then, as in the case of X-rays, the directions satisfying the Bragg relation exhibit a circular symmetry with respect to the direction of the incident beam. For various interplanar spacings a family of cones is obtained, having the tip at the point of contact between the electron beam and the specimen. The intersection of these cones with a plane surface generates a characteristic diffraction pattern, in the form of concentric rings (Figure 9.23 left).

As the angle 2θ is small, the radius of a diffraction ring may be written approximately $R = 2\theta L$, where L is the distance from the specimen to the photographic plate. Replacing the value for the angle θ in the preceding relation by the simplified equation (9.16) a simple interdependence is obtained between the interplanar spacing, the diffraction ring radius R and the instrument characteristic L

$$d_{(h\,k\,l)} = \frac{\lambda L}{R}. \tag{9.17}$$

Most transmission electron microscopes may be adapted to obtain diffraction patterns by means of two techniques: (a) diffraction of a selected area of the specimen, in which case the surface is placed in the normal position for microscopy and (b) diffraction by a convergent beam, in which the specimen is placed under the lens system.

In the first case when the convergence of the beam is ensured by the lens system of the microscope, relation (9.17) becomes $d=K/R$, where K is determined by the instrument characteristics, i.e. the focal length and wavelength of the electron beam. The mechanism of obtaining diffraction patterns for a selected area is shown in Figure 9.24. By diffraction of the incident beam in the area AB of the specimen, in the focal plane F.O. of the objective O, a diffraction pattern is formed; simultaneously, an intermediate pattern $A'B'$ is formed. This intermediate pattern is further magnified by the intermediate and projector lens. If the current of the intermediate lens is adequately changed, a pattern of the plane F.O. in F.I. is obtained, which is then projected onto a fluorescent screen. Thus, by simply varying the current of the intermediate lens, either an electron image of the area AB or a diffraction pattern from this area is produced on the fluorescent screen. By means of several diaphragms of various sizes which are interposed in the objective image plane, a certain area of the specimen AB can be selected, for which the diffraction pattern is then obtained by a simple concentration. The advantage of this technique is that the selection of the specimen area is carried out on the magnified image. Thus, the most interesting areas may be selected. If the specimen is a single crystal, a diffraction pattern similar to that illustrated in Figure 9.23 (right) is obtained.

With the second technique the specimen is placed under the lens system, which has the advantage of eliminating errors caused by lens aberrations and the dependence of calibration on the focal distance and specimen position. The resolution depends on the ring widths or spot diameters, the quality of the electron source image, and the stability of the whole electromagnetic feed system.

The constant K of the instrument may be determined from the optical system characteristics and the wavelength used. These quantities cannot, however, be measured with sufficient accuracy, neither are they constant in most commonly used instruments. For this reason, a more direct method is used, namely the use of a standard substance of known structure.

Usually, a substance crystallized in the cubic system having a single lattice dimension is used. The successive diameters of the diffraction rings are measured on the diffraction pattern obtained, and these are then used for calculating the best values for K.

From relation (9.9) giving the interplanar spacing for the cubic system and relation (9.17) one obtains

$$h^2 + k^2 + l^2 = \frac{a^2}{K^2}(R)^2, \tag{9.18}$$

where, replacing the diffraction ring diameters, the best value for the quantity a/K is found. The lattice parameter a for the substance used as a standard being known, the value of K is finally determined.

To minimize the calibration error, the grid with the standard specimen which is kept permanently in the microscope, is introduced into the electron beam immediately after the material under examination. Thus, within a short time

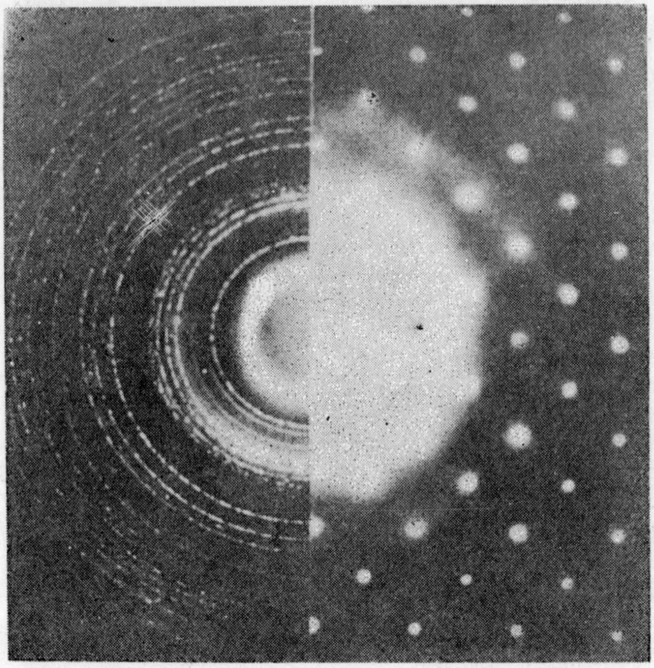

Figure 9.23 Transmission diffraction patterns obtained in an electron microscope.

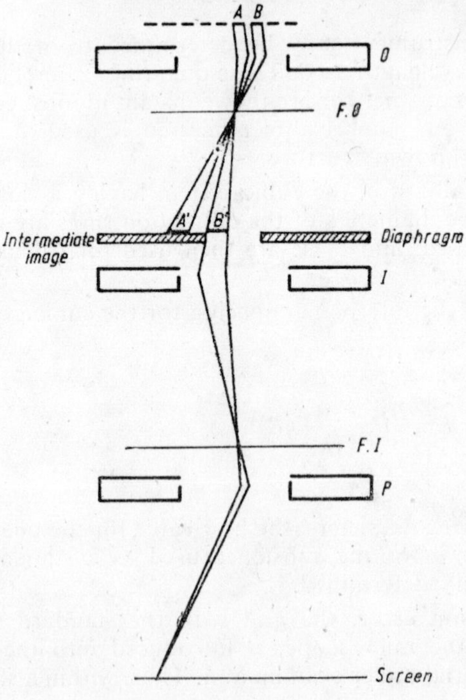

Figure 9.24 Diffraction on selected area.

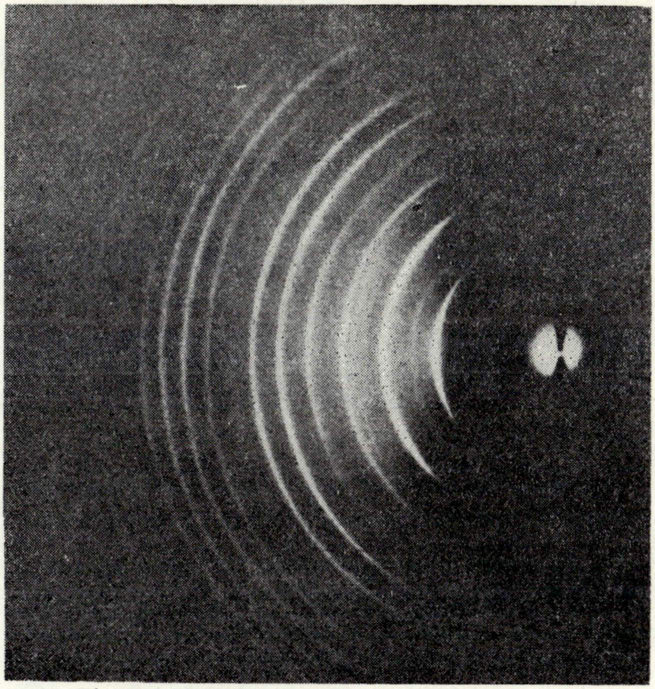

Figure 9.25 Reflection diffraction pattern obtained at glancing incidence.

interval both the standard diffraction pattern and the image of the unknown substance are obtained.

Where thin substance layers are involved the techniques described above achieved diffraction by transmission. However, in most cases when frictional surfaces are examined, a perfect separation of these layers from the metallic support is not possible. For this reason, reflection diffraction is used, which is similar to the method used for X-rays (Figure 9.14). Instead of complete diffraction rings circular arcs are obtained in which the source image is obscured by the specimen (Figure 9.25).

Generally, reflection diffraction patterns are more diffused and their resolution is not as good as that of transmission patterns. Surface roughness affects the resolution of the diffraction pattern with the reflection technique. At high roughness, besides reflection proper, transmission through asperity peaks also occurs. For this reason, where diffraction was used for the examination of thin layers formed on friction pair surfaces only surfaces of low roughness were used.

Applications. Electron diffraction is an efficient technique for determining the composition and structure of surface layers [67]. To determine the composition of a layer, the diameter and intensity of the diffraction lines obtained are measured. By means of the instrument constant the corresponding interplanar spacings are determined. Based on these values one attempts to identify the substance under examination by comparing these values with the values of several known substances.

Sometimes it is useful to have diffraction patterns of some common substances for comparative purposes. For instance, a standard image of α-Fe can be obtained by using a steel surface polished with abrasive paper under a benzene layer to avoid oxidation. For Fe_2O_3 it is necessary to heat the steel surface in air until the interference colours occur while for Fe_3O_4 the heating should be carried out in vacuum. To obtain chloride or sulphide layers the surfaces are heated in HCl or an H_2S atmosphere, respectively.

An essential condition for obtaining standard diffraction patterns is effective surface cleaning. This is especially important for surfaces which have operated in oil. Due to the short path of the electrons any trace of oil, even several molecular layers, will conceal the effect of the layer under investigation.

For organic layers, surface preparation for diffraction is extremely difficult due to their solubility in ordinary solvents.

As X-rays, electron diffraction provides information concerning layer structure and the degree of crystallinity [41]. In the case of a preferred orientation of the crystallites more intense portions will occur around the circumference of the diffraction rings. Depending upon the position of these maxima with respect to the supporting surface the preferred orientation direction can be deduced.

An important result obtained by electron diffraction is the estimation of the mechanism of friction of lamellar solid lubricants (graphite and molybdenum disulphide).

Jenkins suggested that there is a tendency for the (0002) planes of graphite crystals to lie parallel to the surface. Subsequently, when the mechanism was studied in detail it was found that the same planes may lie at an angle δ to the surface. This angle is expressed by the relation $\mathrm{tg}\,\delta = \mu$, where μ is the coefficient of friction determined experimentally [140].

Recent investigation of graphite films formed on copper surfaces led to the crystal pair assumption [143]. In the first stage of operation the irregularities

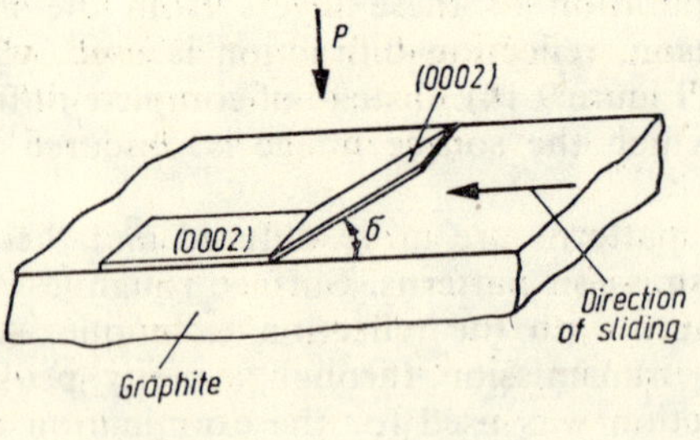

Figure 9.26 Mechanism of action of graphite. The lamella pairs hypothesis.

on the copper surface are filled with randomly-oriented crystallites. Then, an orientation of these crystallites as well as of those in the brush surface occurs so that the (0002) plane is parallel to the direction of motion. Some of these crystallites join in pairs to form an assembly in which one of the crystallites is parallel to the surface and the other makes an angle of several degrees to a maximum value of 22° (Figure 9.26).

Instead of the simple relation tg $\delta = \mu$ which does not hold in the case when $\delta = 0$ the following relation is suggested

$$\mu = \mu_p + \mu\delta, \tag{9.19}$$

where μ_p is the coefficient of friction due to those crystallites which are permanently parallel to the surface.

9.5 The radioactive tracer method

9.5.1 Fundamentals of the method

Between two isotopes of the same element only a difference in atomic mass exists, the electronic shell structure being the same in both cases. As the chemical properties depend on the configuration of the (outer) electronic shells, it follows that two different isotopes of the same element will behave similarly from a chemical point of view. The mass effect which occurs in the phenomena involving transport of material is negligible in most technical processes.

All elements have amongst their natural or artificial isotopes radioactive isotopes which can be detected readily due to the radiations they emit as a consequence of disintegration. The great resemblance of the physico-chemical properties of the isotopes of the same element, together with the possibility of detecting radioactive isotopes is the basis of the radioactive tracer technique. By following the evolution of a radioactive isotope in a process conclusions may be drawn concerning the behaviour of the stable isotope which represents the main part of the isotope mixture.

In a specimen if the total number of atoms of a species is N_i and the number of radioactive atoms of the same species is N_i^*, then, based on the identical behaviour of the two isotopes in any process

$$\frac{N_i^*}{N_i} = \text{ct.} \tag{9.20}$$

This ratio is known and by measuring N_i^* before and after the process N_i can be determined.

Radioactive Disintegration. Detection of radioactive isotopes is effected by recording the radiations emitted at the radioactive disintegration. Any radioactive isotope undergoes a spontaneous disintegration process in which corpuscular or/and electromagnetic radiations are emitted. The composition of the radiation emitted may be represented schematically by the disintegration diagram. Figure 9.27 shows the disintegration diagram of the ^{59}Fe isotope from which it follows that the radiation emitted consists in the main of β radiation of 0.27 MeV and 0.46 MeV and radiation γ of 1.098 MeV and 1.289 MeV.

The quantity of radioactive atoms in a specimen decreases by disintegration. If at a given moment the number of radioactive atoms in a specimen is N_0^*, then N^* atoms will exist after a time t

$$N^* = N_0^* e^{-\lambda t}, \tag{9.21}$$

where λ is the disintegration constant.

To characterize the disintegration rate the half-life value, i.e. the time in which $N^* = N_0^*/2$ is also used. Between the half-life value $T_{1/2}$ and the disintegration constant a simple relationship exists $T_{1/2} = 0.693/\lambda$.

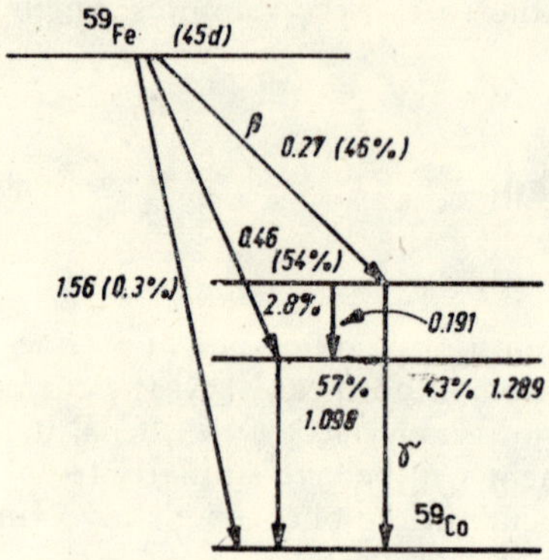

Figure 9.27 ^{59}Fe disintegration diagram (energies in MeV).

The number of atoms disintegrating in unit time is called the activity Λ of the specimen

$$\Lambda \equiv -\frac{dN^*}{dt} = \lambda N^*. \tag{9.22}$$

By measuring the number of disintegrations in unit time the number of radioactive atoms existing in the specimen may be deduced.

Measurement of the radiations emitted by a radioactive substance is carried out by means of a radiation transducer and an ancillary electronic equipment. In most popular transducers the radiation flux is transformed into a succession of electrical impulses which are counted.

Between the counting rate R (number of impulses counted in unit time) of a transducer-counter assembly and the activity of a radioactive specimen the dependence is not simple. As seen, a disintegrating nucleus can emit a number of particles of various types and energies. Usually due to the measuring geometry, the beam emitted is only partly intercepted by the transducer. Also, the transducer does not exhibit an equal efficiency with all types (and energies) of measured radiation. Under constant measuring conditions the counting rate will however be proportional to the specimen activity

$$R = c \cdot \Lambda. \tag{9.23}$$

The constant c takes account of the disintegration scheme factor, of the solid angle under which the transducer is viewed by the source, and the efficiency of the transducer with the various types of radiations, etc. Although detailed determinations exist for each of the components of the constant c, the relationship between R and Λ is established exactly by calibration.

Measuring errors. The statistical character of the disintegration process is also reflected in the counting rate R. If k impulses are counted in time t, then the standard deviation σ_k of counting is

$$\sigma_k = \sqrt{k}, \tag{9.24}$$

while the standard deviation of the counting rate $R = k/t$ is σ_R

$$\sigma_R = \sqrt{\frac{R}{t}}. \tag{9.25}$$

If R is determined as the mean of several measurements, then t is the total time.

The standard deviation of the counting rate σ_R indicates the range $(R-\sigma_R) \div (R+\sigma_R)$ within which the results of any measurement carried out under the same conditions will lie with a probability of 68%.

Besides the radiation coming from the source a background radiation is also measured (for instance, the permanent cosmic rays) and the counting rate corresponding to the specimen is

$$R = Q - F, \tag{9.26}$$

where

Q is total counting rate (imp/min),

F — background counting rate (imp/min).

As both radiation sources are random the error as far as σ_R is concerned is

$$\sigma_R^2 = \sigma_Q^2 + \sigma_F^2. \tag{9.27}$$

If it is assumed that the counting rate t of the background and specimen in the presence of the background is the same, then the standard error concerning R is

$$\sigma_R = \sqrt{\frac{Q+F}{t}}, \tag{9.28}$$

whereas the relative error ε_R

$$\varepsilon_R = \frac{\sigma_R}{R} = \frac{1}{\sqrt{t}} \frac{\sqrt{Q+F}}{Q-F}. \tag{9.29}$$

From relation (9.29) the counting time required for a certain relative error ε_R is

$$t = \frac{1}{\varepsilon_R^2} \frac{Q+F}{Q-F}. \tag{9.30}$$

Radiation Transducers. Gas counters, scintillation counters and photographic emulsion are the most frequently used radiation transducers.

Gas counters, like the Geiger-Müller counter, consist of two electrodes between which a potential difference exists. These electrodes are enclosed in a volume of gas of known composition. The ionization process initiated by the passage of a charged particle through the transducer active volume causes an electrical impulse in the counter supply circuit. This impulse is processed in the electronic circuit and recorded. As not all particles reaching the transducer active volume produce an impulse, its efficiency, determined as the ratio of the number of impulses to the number of particles which produced them, is always less than unity. The efficiency of a G.M. counter depends on the type and energy of the radiation. For β radiation the efficiency of the G.M. counter approaches unity provided the particles pass through the counter wall. For γ radiation efficiency falls to several percent.

Scintillation counters use the property of several materials to transform the energy of γ radiation into light energy. The photons emitted by the scintillator are amplified by a photomultiplier and transformed into electrical impulses. The efficiency of scintillation counters for γ radiation is much higher than that of G.M. counters due primarily to the much greater density of the substance in the active volume. The material is either a crystal (for instance, NaI activated by thallium) or a liquid. By using perforated crystals one can achieve a measuring geometry close to 4π, for a small volume of material. In this case the total efficiency goes up to 80—90%.

Photographic film transducers are based on ionizing radiations activating silver bromide emulsion. As the transformation is not reversible, this technique allows the recording of a quantity proportional to the total amount of radiation penetrating the emulsion. The quantity measured is the blackening of the film. The advantage of this method of radiation detection is that it can establish the space or plane distribution of the radioactive source.

Radioisotope Production. Due to the possibility to obtain artificial isotopes, the radioactive tracer technique was developed. The main source of radioactive isotopes is the nuclear reactor in which a nuclear reaction takes place between neutrons and the nuclei of the target substance. By capturing a thermal neutron an isotope of the target element is obtained and a photon γ is eliminated. For example, to obtain the radioisotope ^{60}Co with thermal neutrons the reaction

$$^{59}\mathrm{Co} + {_0}n^1 \rightarrow {^{60}\mathrm{Co}} + \gamma$$

takes place.

This reaction may be written briefly as $^{59}\mathrm{Co}(n, \gamma)\,^{60}\mathrm{Co}$.

The probability that a certain nucleus will react with the projectile particle is described by the concept of cross section. The cross section σ for a certain nuclear reaction is given by the area of a disc perpendicular to the direction of the beam. The nuclear reaction takes place when the beam impinges on the disc. The cross section may be larger or smaller than the geometrical cross section of the nucleus of the target.

The cross section concept is useful for the calculation of the amount of radioisotopes produced in a nuclear reaction. The attenuation of a particle beam, e.g. of a neutron beam, as a consequence of the nuclear reaction is

$$I = I_0 \, e^{-N\sigma x}, \tag{9.31}$$

where

I, I_0 are the beam intensities, before and after passing through the target (neutrons/cm^2),

N — number of atoms per cm^3,

x — target thickness (cm),

σ — cross section (cm^2).

The number of radioactive atoms formed is

$$I_0 - I = I_0(1 - e^{-N\sigma x}). \tag{9.32}$$

The cross section unit is the barn; 1 barn $= 10^{-24}$ cm^2.

Disintegration and formation of radioisotopes occur simultaneously in the target; thus at a certain moment an equilibrium concentration is established which depends on the neutron flow rate, the cross section and the half value life.

9.5.2 Application of radioisotopes to the study of tribological processes

The main process to which relation (9.20) refers and which accounts for the usefulness of the radioactive tracer technique is the study of processes involving mass transfer. In tribology, radioactive tracers have been applied to the study of the following phenomena:

(a) surface wear with the formation of wear debris,

(b) transfer of material between surfaces,

(c) interaction between the surface and lubricant.

Surface wear. To determine the wear of a component by means of radioactive tracers it is necessary to have a uniform distribution of the isotope over the whole volume of the component or at least in a sufficiently small volume of material. This is necessary as alloys have a microstructure which may affect the distribution of the radioisotope. If this condition is obeyed, relation (9.20) is also satisfied for the wear process.

A typical scheme for the study of wear by means of radioisotopes is shown in Figure 9.28. The friction pair, consisting of parts A and B, is immersed or is in contact with a volume of lubricant V_T. One or both parts, A or B, contain the radioactive tracer. The wear debris which is produced accumulates in the oil volume V_T. A fraction V_M of the total oil quantity is continuously or periodically measured by the transducer. From the activity of the volume measured V_M the amount of worn material is deduced.

An important stage in the designing of an experiment is the calculation of the activity required to ensure a certain measuring sensitivity and accuracy.

For calculating the activity of the part A (assuming that only the wear of this part is assessed) the following data are required:

— the minimum wear volume required to be detected with a specific accuracy (Δv),

— total wear volume which is to be measured (Δv_{max}),

— measuring sensitivity $S = \dfrac{R}{\Delta v}$,

— maximum permissible standard error ε_R.

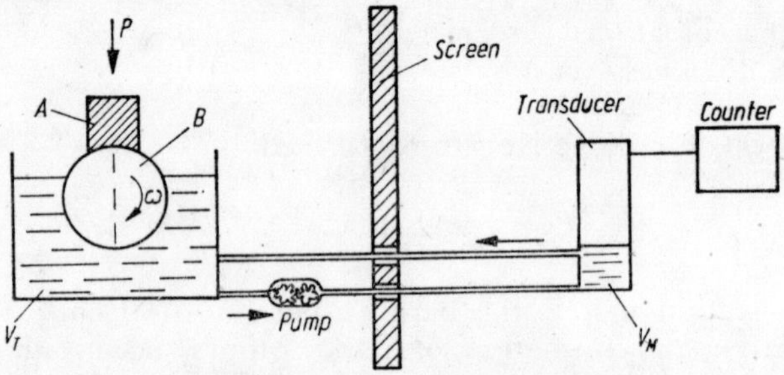

Figure 9.28 Apparatus for the study of wear by isotopes.

The minimum volume Δv determines, by means of the surface which is worn away, the minimum measurable thickness. This computation assumes uniform wear over the whole surface.

The total volume which is to be measured is important as it determines the background in which the activity corresponding to the volume Δv is measured. As the technique used in Figure 9.28 is cumulative, the wear debris present in the lubricant constitutes the background against which the activity corresponding to the volume Δv is measured.

If at a given moment the counting rate Q_i is recorded and by increase of the wear volume in the lubricant by Δv the counting rate increases to Q_j, then

$$Q_j - Q_i = R, \tag{9.33}$$

where Q_i is the background for determining R.

The counting rate at a given moment is

$$Q = \frac{V_M}{V_T} \varepsilon \, v \Lambda_s \rho, \tag{9.34}$$

where

$\dfrac{V_M}{V_T}$ is the ratio of the measured volume to the total volume of lubricant,

ε — transducer overall measuring efficiency (including the measuring geometry and efficiency of the transducer with respect to the radiation measured),

v — volume of worn material in the lubricant (assumed to be uniformly distributed),

Λ_s — specific activity (with respect to the mass) of part A,

ρ — density of the part material.

The sensitivity $S = R/\Delta v$, together with the minimum volume which is to be measured Δv determines the value of R. With this value the necessary specific activity is calculated by using relation (9.34).

Then, Q_{max} is calculated by substituting into relation (9.34) the previously calculated values v_{max} and A_s. The necessary measuring time is checked so that the error concerning R is not larger than ε_R. The computation is performed in the most unfavourable situation, that is at the last measurement when the background was maximum. Substituting in relation (9.30) $F \simeq Q = Q_{max}$ and $Q - F = R$ the necessary measuring time is obtained. If the resulting measuring time is too great, it is necessary to decrease the background, that is the maximum value of the wear products in the lubricant. The value of the ratio R/Q_{max} may be reduced without altering the development of the experiment by changing the lubricant. In other words, if a total wear v_{max} is required, and the resulting time is too great, this volume is divided into several fractions by changing the oil. In this case Q_{max} becomes $Q_{max} = Q_{max}/n$ where n is the number of oil changes.

Calculation of the protection against radiations is performed on the basis of the total activity of the part A

$$A = gA_s, \tag{9.35}$$

where g is the weight of the part.

A high total activity causes problems in ensuring protection in service. Also, the specific activity can not be reduced as it determines indirectly the measuring sensitivity.

In the case of a uniform distribution of the radioisotope in the part volume a reduction in the total activity without diminishing the sensitivity can be achieved by:

— increasing the ratio V_M/V_T, especially by reducing V_T,
— increasing the overall efficiency ε of the transducer.

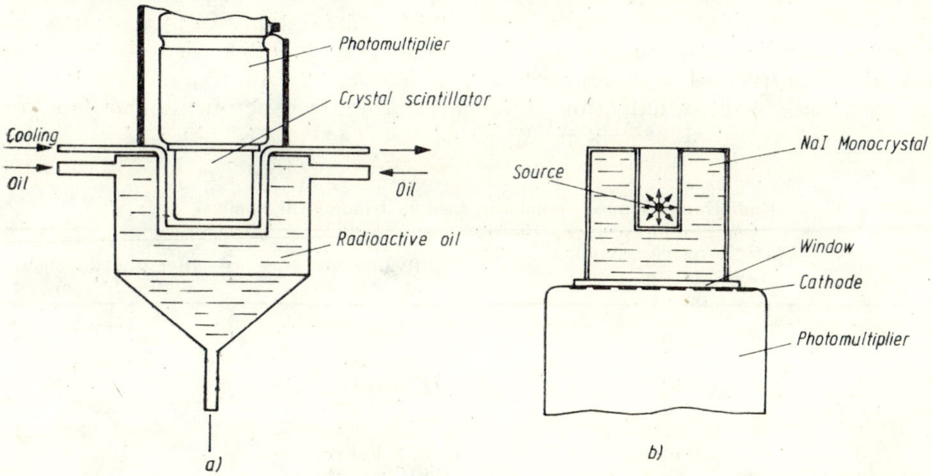

Figure 9.29 Radiation transducers with scintillator and photo multiplier: *a*, with lubricant circulation, *b*, with perforated crystal.

The above quantities are not completely independent. Generally, an improvement in overall efficiency means an increase in the cost of the transducer. For improved efficiency scintillation transducers are currently used. Figure 9.29 *a* shows an instrument for the continuous measurement of lubricant activity, and Figure 9.29 *b* shows a specimen measuring device which uses a perforated scintillator crystal. The increase in geometrical efficiency in the latter case is achieved however to the detriment of the ratio V_M/V_T.

The sensitivity which can be achieved in determining wear on a real aggregate, without taking any special precaution against radiations is of the order of 10^{-6} g wear/h. For laboratory apparatus the sensitivity may be even higher.

In some cases, in laboratory experiments the whole volume of lubricant was measured when $V_M/V_T = 1$, i.e. it is a maximum. This technique has also the advantage of eliminating error caused by sedimentation of wear particles in the measuring circuit [135].

Activation Techniques. The following activation techniques are used:

(a) integral part activation,

(b) introduction of the radioisotope during casting of the specimen,

(c) deposition of the radioisotope on the friction surface by: diffusion, electrolytical deposition and insertion of a point of reference,

(d) thin layer activation.

The simplest and most reliable method is the integral activation of part *A* (Figure 9.28) in the nuclear reactor. Its advantages include a uniform distribution of the radioisotope in the whole volume of the part; the handling of the radioactive material is minimized as the part is irradiated following the final processing stage; the properties of the material of the part are not affected in any way.

Calculation of the activity of the part should take into account the composition of the alloy from which the part is manufactured. After irradiation, a mixture of radioisotopes is obtained, the concentration of each being a function of the element concentration, its cross section, its half-life value, irradiation time and deactivation time. The computation is simplified by neglecting short-lived isotopes. Irrespective of cross section, their contribution to total activation becomes negligible after several half-life periods (Figure 9.31). Also, the presence of several short-lived isotopes means that frequent calibration is required as a consequence of the deactivation of the part during the experiment. The standard

Table 9.3

Radioisotopes most frequently used in tribological studies

Isotope	Max(*) specific activity, mCi/g	Half-value life	Specific activity after one day irradiation (*), mCi/g
^{32}P	110	14.3 days	5.3
^{35}S	5.9	87.1 days	0.047
^{50}Cr	160	27.8 days	4
^{59}Fe	0.37	45.1 days	0.0055
^{60}Co	6 400	5.27 years	2.3
^{63}Cu	510	12.8 hours	—
^{65}Zn	64	250 days	0.18
^{99}Mo	18	67 hours	—
^{110}Ag	160	270 days	0.4
^{114}In	340	49 days	4.8
^{192}Ir	34 000	74.37 days	320

(*) Activity obtained by irradiation of the natural mixture of isotopes with a neutron flux of 10^{12} part/cm² used for determining the constant from relation (9.23) is also used for establishing the deactivation correction.

The main characteristics of radioisotopes most frequently used in tribology, obtained by neutron activation, are summarized in Table 9.3. Besides the half-life value the table also gives the activity after one day of irradiation and the maximum activity obtained in a neutron flux of 10^{12} part/cm^2 s.

Sometimes, it is found that for the available flux the composition of the irradiated alloy part does not reach the specific activity required for a certain sensitivity. In this case, an element with a large cross section and a convenient half-life value is introduced into the alloy part. An additional condition is the complete miscibility of the tracer element in the material. For alloys, the element should be soluble in the main constituent or disperse completely through the alloy volume. Tracers are usually used when bronze or plain carbon steel are activated. The main components of these alloys Cu, Sn or Fe have too short a half-life or do not lead to high specific activities. To increase the specific activity of bronzes, indium can be used or, for cast irons and plain carbon steels iridium can be used, both in concentrations of 0.5% or less. No important change in the properties of the above alloys was reported as a consequence of the use of iridium and indium in low concentrations [134, 135].

The radioisotope is introduced during the casting process of the part only if its size does not allow irradiation in the nuclear reactor. This method has the disadvantage that the whole operation of processing of the part must be carried out with radioactive material.

The method of part activation by deposition of the radioisotope on the frictional surface of the part may be carried out in three ways (Figure 9.30).

A radioactive isotope of a component in the alloy is introduced by diffusion into the surface. The disadvantage of this method is a non-uniform concentration of the radioisotope in the bulk of the material and the change of chemical composition, and so of the surface properties.

The technique of electrolytic deposition may be applied only when this is also used in the normal manufacturing process. In this case, the electrolyte is radioactive while the part is inactive.

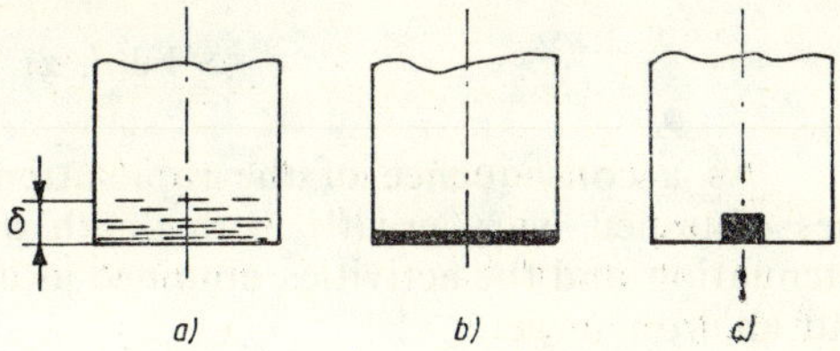

Figure 9.30 Techniques for concentrating the radioisotope at the part surface:
a, introduction of the radioisotope by diffusion; *b*, electrolytical deposition; *c*, radioactive pin.

A variation of this technique is the deposition of an inactive layer on an inactive support and irradiation of the whole part. By proper selection of the support as well as of the irradiation and deposition times the activity can be concentrated mainly on the surface of the part. Figure 9.31 shows the variation of the activity at activation and deactivation for a bronze support and a silver layer.

Another method for concentrating the activity on the frictional surface of the part is by the use of insertions (Figure 9.30 *c*). The pin introduced may have a high specific activity to compensate for its small area. A special problem raised by this method is to provide conditions so that wear of the reference material follows the wear of the surface and its presence does not change the local lubricating conditions. In order to do so, it is necessary to consider pin hardness, its stability in the groove, etc.

Thin Layer Activation. A highly efficient technique for reducing the total radioactivity without decrease in the specific activity is thin layer activation [196, 197]. In this case radioactive isotopes are obtained by nuclear reactions

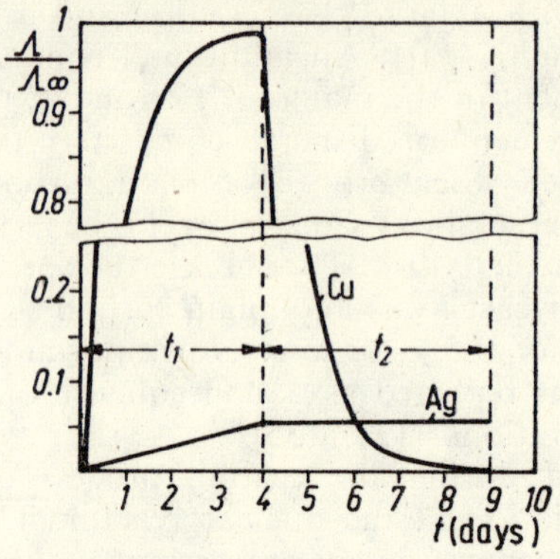

Figure 9.31 Activation—disactivation curve for ^{64}Cu and ^{110}Ag.

between high energy charged particles from an accelerator and the atoms of the surface to be irradiated. As distinct from neutrons the path of the beam of heavy, charged particles is much shorter (of the order of several fractions of a millimeter for attenuation in iron). If the radiation beam has sufficient energy, it causes nuclear reactions to occur which may also lead to radioactive isotopes. If a deuteron beam is used on an iron target, the following reactions leading to radioactive isotopes take place:

$$^{56}\mathrm{Fe}(d, 2n)\ ^{56}\mathrm{Co}$$

$$^{56}\mathrm{Fe}(d, n)\ ^{57}\mathrm{Co}$$

$$^{54}\mathrm{Fe}(d, \alpha)\ ^{52}\mathrm{Mn}.$$

As a consequence of the rapid attenuation of the beam, the specific activities obtained vary greatly with depth of the material. Figure 9.32 shows the attenuation and the activities obtained as a function of depth for a deuteron beam and an iron target.

To correlate activity with the volume of worn material, an activity vs. depth curve is required. Generally, this curve has a complicated shape, which makes determination of the worn volume difficult. By selecting a certain beam energy a specific activity which is constant with depth may be obtained for limited material thicknesses. This situation allows a linear variation of total activity as a function of depth (Figure 9.33). The linear range is arranged so that it begins right at the material surface.

The main advantage of this method is the provision of specific activities comparable with those obtained by neutron irradiation, the total activity of the part being low. Also, the size and position of the active surface may be specifically chosen.

Determination of the volume of worn material is performed either by measuring the activity of the lubricant or by measuring the total activity of the

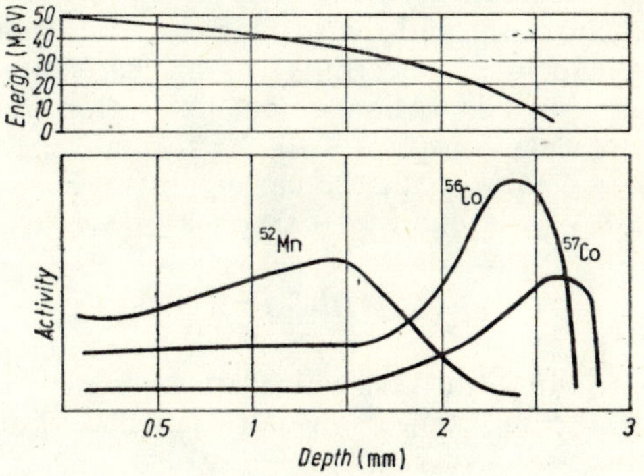

Figure 9.32 Deuteron energy and activity versus depth for an iron target [196].

irradiated part. The latter method, which is possible only as a consequence of a significant reduction in the total activity of the part, allows determination of the wear of machine parts in service. The activity of the part the wear of which

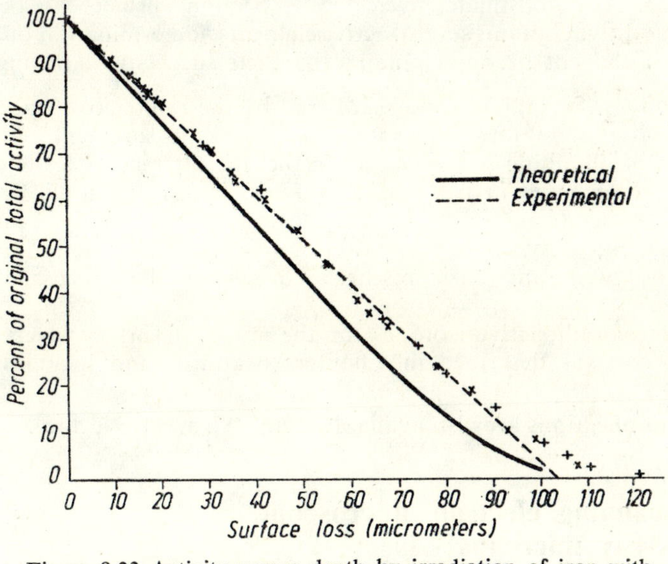

Figure 9.33 Activity versus depth by irradiation of iron with deuterons of 9.2 MeV energy [197].

is examined is continuously measured by means of an external counter without dismantling the machine. A sensitivity better than 1 μm wear for a total thickness of 100 μm was obtained. The sensitivity may be improved by oblique irradiation, which determines a decrease in the thickness of the activated layer.

Interaction between Lubricant and Surface Material. To investigate the interaction between the lubricant or additives and metallic surfaces synthesized radioactive additives are used. Thus, one or several atoms in the molecule can be labelled. A knowledge of the position of the labelled atom contributes to the elucidation of the mechanism of interaction between the molecule of additive or lubricant and the surface material. Inadequate labelling may lead to erroneous conclusions concerning the mechanism of interaction.

In adsorption or chemisorption studies where it is known that the whole molecule contributes to the formation of the layer, any atom may be labelled. For a fatty acid, the carbon in the active group or the carbon at the opposite end of the chain of carbon atoms is usually labelled

$$H_3C^* - CH_2 - \ldots\ldots\ldots\ldots CH_2 - \overset{*}{C}OOH.$$

In additives, which are known to form reaction layers, the active element of the molecule is usually labelled. For instance, by labelling phosphorus in tricresylphosphate

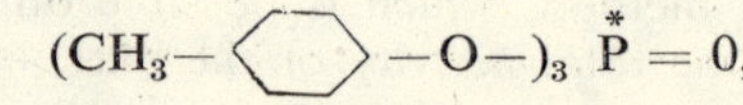

the formation of a phosphate layer on the friction surface can be examined.

If the additive contains several active elements, determination of the importance of each element becomes possible by their successive labelling.

Applications. Several advantages offered by the radioactive tracer technique (high sensitivity, possibility of continuous wear measurement, etc.) together with the ease of obtaining radioisotopes in the nuclear reactor leads to this technique being widely used in tribological studies. Important applications include engine wear studies (ring-jacket assembly, cam-tappet assembly, bearings, etc.); gear wear studies; study of the formation of adsorbed and reaction layers; study of cutting tool wear and basic research concerning the process of boundary lubrication.

The use of radioactive isotopes for the study of surface layers formed on friction surfaces was described in Chapters 6 and 7, and includes a relevant bibliography.

Other publications are also available [176, 48], and [134, 135].

9.6 The scanning electron microscope and X-ray microanalysis

In the last decade electron probe microanalysis for surface analysis has been developed. In this technique the surface to be examined is scanned by an electron beam of small diameter (Figure 9.34). Interaction between electron beam and the surface results in backscattered electrons, secondary electrons and characteristic X-radiation which may be used for the analysis of the composition or topography of the surface. A feature of electron probe X-ray microanalysis is scanning of the specimen and the subsequent correlation of information obtained by the analysis with the electron image of the surface.

The scanning electron microscope, whose simplified block diagram is shown in Figure 9.35, usually uses secondary electron emission to obtain an image of the surface. The electron beam, accelerable through a potential difference

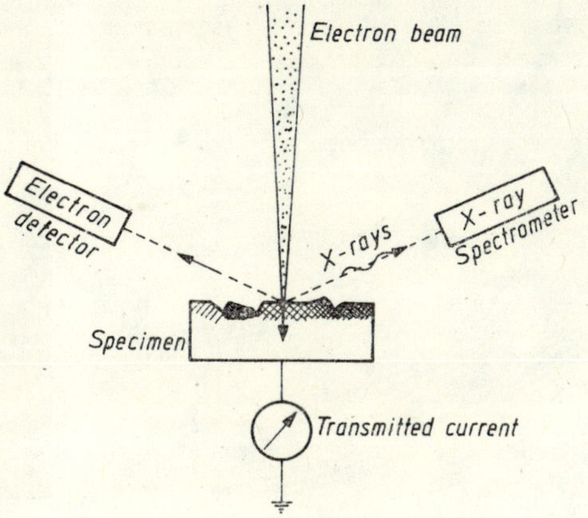

Figure 9.34 Micro-probe schematic diagram.

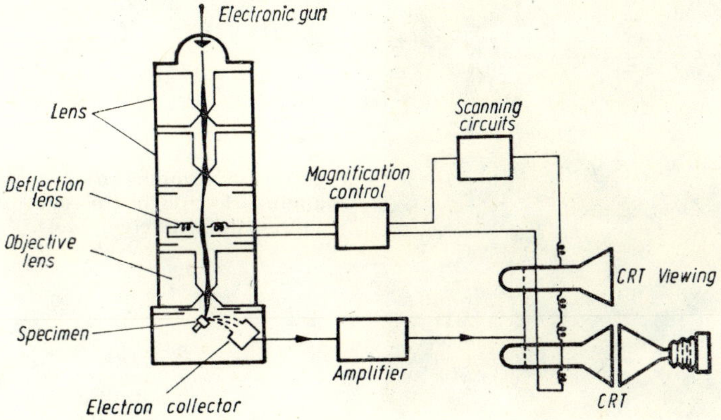

Figure 9.35 Scanning electron microscope schematic diagram.

up to 30 keV, is focused by two condenser lenses. The position of the spot on the specimen surface is controlled by a system of deflection coils interposed between the condenser lens and the objective lens which ensures the final focussing. The deflection system allows surface examination by scanning similar to the technique used in television.

The secondary electrons emitted by the specimen are collected by a scintillator-multiplier transducer. This gives an electrical signal proportional to the intensity of the secondary electron flow, which is used, after suitable amplification, to modulate the intensity of a cathode ray tube. By a displacement in synchronism of the two electron beams (from the microscope and the cathode ray tube) an optical picture of the surface being examined is produced. It should be noted that, unlike the other microscopes, this instrument does not provide a direct image of the surface; magnification is obtained only by the different scanning

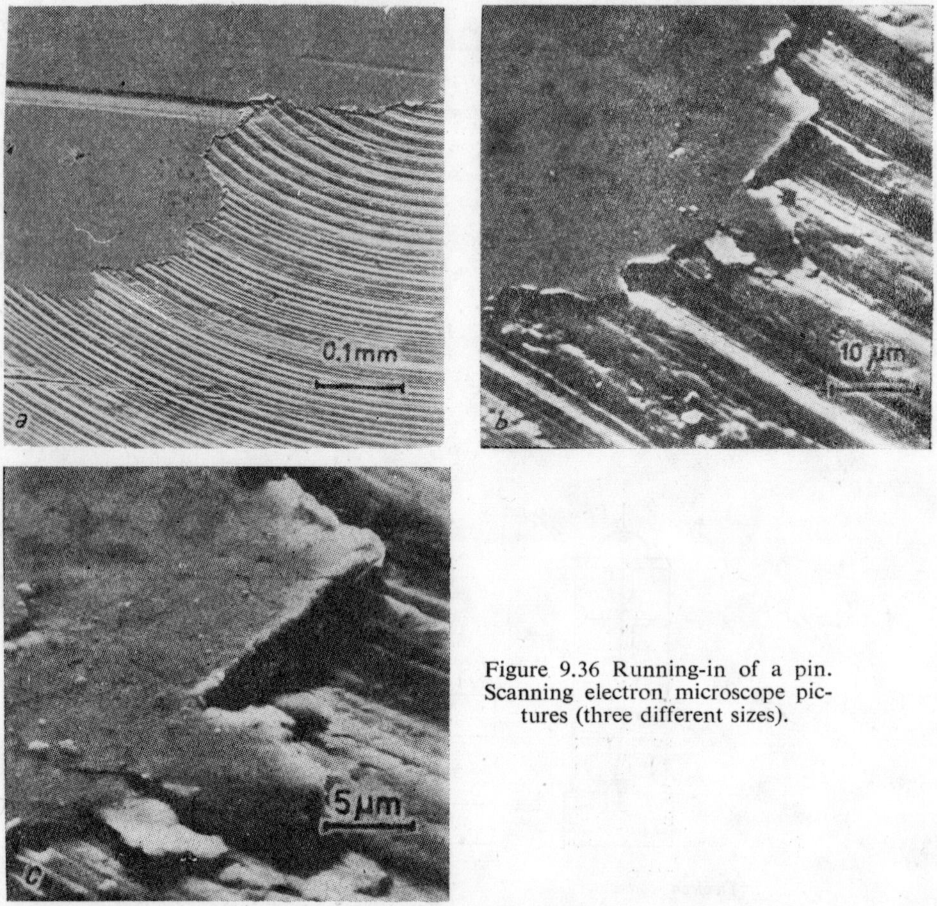

Figure 9.36 Running-in of a pin. Scanning electron microscope pictures (three different sizes).

amplitudes of the beam of analyser electrons compared with that of the cathode ray tube. The final image is formed point by point and the resolution depends on the diameter of the spot of specimen electrons. At present, diameters below 1 µm are possible, which leads to practical resolutions under 1 µm. The magnification is variable from 20 × to 50000 ×. The depth of field is about 1000 × greater than of optical microscopes.

An important advantage of the scanning electron microscope is direct specimen observation without previous preparation of metallic surfaces. Generally, the specimen size is large (e.g. Ø 12). Also, it may be moved in all three directions and rotated around a vertical axis. Another advantage is that different magnifications are possible without change of adjustment.

The above advantages make the scanning electron microscope an extremely useful instrument for the study of worn surfaces. The high resolution and great depth of focus is useful for the study of changes in surface topography by wear processes.

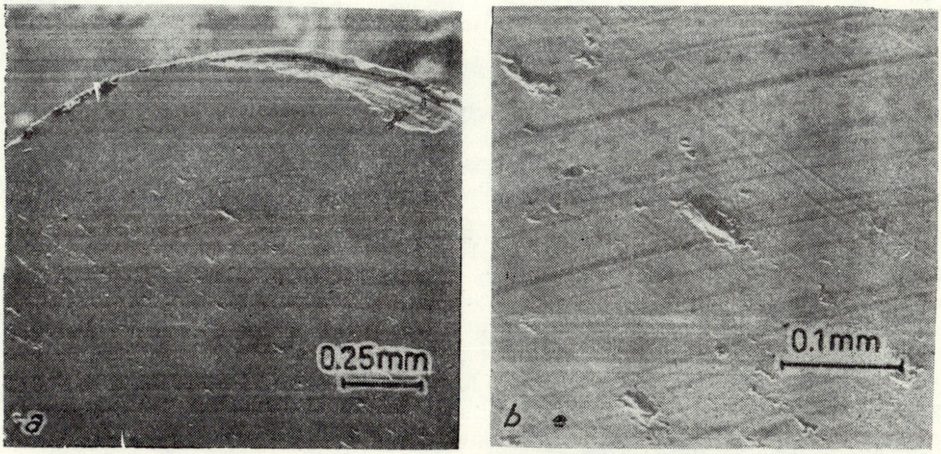

Figure 9.37 Surface defect. Scanning electron microscope pictures (three different sizes).

X-Ray Microanalysis. Analysis of the characteristic X-radiation emitted by the material with which the electron beam has interacted permits a qualitative and quantitative determination of the component elements. Due to the small beam diameter, point analysis is conferred to a volume of material of only several μm^3. The signal obtained from an X-ray spectrometer is either transmitted to a recorder or amplified to modulate the intensity of a cathode ray tube.

The elemental analysis may be carried out in two ways:

(a) the whole X-ray spectrum emitted by a point is recorded (point elemental analysis). The point to be examined is selected by an optical microscope which permits observation and positioning of the specimen in the vacuum chamber.

(b) the X-ray spectrometer is adjusted to the characteristic line of an element; the surface is then scanned and a result is obtained in the form of an image on the cathode ray tube in which the intensity of each point (or number of points) is approximately proportional to the element concentration. Usually, microanalysers are provided with two or three spectrometers, to allow simultaneous analysis of several elements.

Problems encountered in conventional X-ray quantitative analysis (Chapter 9.1) are also inherent in microanalysis. Here, too, it is necessary to use standards, correction curves for self-adsorption, etc., to obtain quantitative results with a minimum of error. However, the possibilities offered by microanalysis for the study of microscopic structures ensure continuing research in its development. Whilst scanning microscopes are specialized instruments, microanalysers are more versatile and can be used as scanning microscopes.

Applications. The scanning electron microscope is being increasingly used to advantage in the study of the process of solid surface wear. The main qualities of this instrument, namely high contrast (due to the "oblique" illumination), a great depth of field and large magnification range are illustrated in Figures 9.36—9.37. In addition to a rough turned area, a very flat run-in area can also be observed. The manner in which this surface forms is questionable. In any case, besides possible classical wear by particle separation, it appears that plastic deformation of the roughness also occurs, which causes flow and a flattening of the surface material. The manner in which this process develops may be examined at the edge of the flat area especially at higher magnification.

Even after the whole surface of the pin has been smoothed the effect of plastic deformation can be observed by examination of the small surface defects at a suitably high magnification (Figure 9.37). Obviously, these defects did not form on the flat surface but represent unfilled voids. It appears that under conditions of mixed lubrication preventing frequent contact between the surfaces changes of microtopography are effected by plastic deformation and wear occurs by material removal.

Electron probe X-ray microanalysis for determination of the composition of the surface layer is a relatively new technique [173, 141] (Figure 7.3) which will be more widely used as new more effective instruments become available.

Bibliography

1. ADAMSON, A. W., *Physical Chemistry of Surfaces*, Interscience Publishers, 1967.
2. ADLER, K., SCHOON, TH. G., *Untersuchungen zur Kentnis aer Schmier- und Vershleiß-vorgänge im Gebiet der Grenzreibung*, Schmiertechnik + Tribologie, **15**, 257–271 (1968).
3. ALBERTSON, C. E., *The mechanism of antisquawk additive behavior in automatic transmission fluids*, ASLE Trans., **6**, 300–315 (1963).
4. ALLUM, K. G., FORD, J. F., *The influence of chemical structure on the load-carrying properties of certain organosulphur compounds*, J. Inst. Pet., **51**, 145–161 (1965).
5. ALLUM, K. G., FORBES, E. S., *The load carrying properties of organic sulphur compounds. Part II. The influence of chemical structure on the anti-wear properties of organic disulphides*, J. Inst. Pet., **53**, 173–185 (1967).
6. ALLUM, K. G., FORBES, E. S., *The load carrying mechanism of organic sulphur compounds. Application of electron probe microanalysis*, ASLE Trans., **11**, 162–175 (1968).
7. ALLUM, K. G., FORBES, E. S., *The load carrying properties of metal diakyl dithiophosphates: the effect of chemical structure*, Proc. Inst. Mech. Eng., 1968–1969, **183**, Pt. 3P 7–14.
8. ARCHARD, J. F., HIRST, W., *The wear of metals under unlubricated conditions*, Proc. Roy Soc., ser. A, **236**, 397–410 (1956).
9. ARCHARD, J. F., HIRST, W., *An examination of a mild wear process*, Proc. Roy. Soc., ser. A, **238**, 515–528 (1956).
10. ARCHARD, J. F., *Elastic deformation and law of friction*, Proc. Roy. Soc., ser. A, **243**, 190–205 (1957).
11. ARCHARD, J. F., *The temperature of rubbing surfaces*, Wear, **2**, 438–455 (1958/1959).
12. ARDENNE, M. (von), *Tabellen zur angewandten Physik*, Band I, VEB, D.V.W., Berlin, 1962.
13. AKIN, L. S., *Method of selecting an optimum lubricant testing system*, ASLE Trans., **9**, 249–256 (1966).
14. ASKWITH, T. C., CRONCH, R. F., CAMERON, A., *The relationship of molecular chain length of lubricants and additives for optimum effect, and a theory of scuffing*, Symposium on Gear Lubrication, London, 1966, Part II. Theoretical aspects of gear lubrication.
15. ASSMANN, H., *Vergleichende Untersuchungen von Getriebeölen im FZG Stirnrad- und Hypoidgetriebeprüfstand*, VDI-Zeitschrift, **6**, 255–269 (1965).
16. BAIST, W., *Ein neues Gerät zur Untersuchung der Schmierung metallischer Reiboberflächen bei hohem Druck bzw. bei hoher Temperatur*, Erdöl und Kohle. Erdgas. Petrochemie, **22**, 24–29 (1969).
17. BAIST, W., *Untersuchungen über die Schmierung metallisher Reiboberflächen mittels chemisher Reaktionsschichten*, Diss. Tech. Univ. Clausthal, 1971.
18. BALL, R. E., HERFERT, R. E., *Preparation and characterization of a new crystallic form of MoS_2*, J. Am. Chem. Soc., **79**, 3351/3354 (1957).
19. BARCROFT, F. T., *A technique for investigating reactions between E.P. additives and metal surfaces at high temperature*, Wear, **3**, 440–453 (1960).
20. BARCROFT, F. T., DANIEL, S. G., *Extreme-pressure film formation on hypoid gears studies with sulphur-35*, Lubrication and Wear, Third Convention, London, 1965. Proc. Inst. Mech. Eng. (London), 179/3, 265/277 (1964/1965).

21. Bartz, W., Göttner, G. H., Moos, J., *Vergleichende Bewertung von Kraftzeuggetriebe-Schmierölen mit verschiedenen Ölprüfgeräten*, Mineralöl-Technik, **10**, (1965).

22. Bartz, W., *Grundlagen der Getriebeschmierung. Die Maschine-Antriebs- und Getriebetechnik* (1970), nr. 4, 70–74, nr. 5, 35–39.

23. Bartz, W., Schultze, G. H., Göttner, G. H., *Zur Frage der Schmierwirksamkeit von MoS_2 in Motoren- und Getriebeölen*, Erdöl-Erdgas-Zeitschrift, **84**, 16–25 (1967).

24. Bartz, W., *Schmierung mit Molybdändisulfid*, Technische Rundschau, **44**, **45**, **49** (1969).

25. Bartz, W., *Einfluß des Getriebeöles auf Schäden an Zahnrädern*, Antriebstechnik, **9**, *10*, 1–6 (1970).

26. Beerbower, A., Greene, D. F., *The behavior of lubricating oils in inert gas atmospheres*, ASLE Trans., **4**, 87–96 (1961).

27. Benedict, G. H., Kelley, B. W., *Instantaneous coefficients of gear tooth friction*, ASLE Trans., **4**, 59–70 (1961).

28. Bennett, P. A., Kabel, R. A., *The anti-scuff performance of metal-organodithiophosphate additives*, Lubr. Eng., **19**, 365–370 (1963).

29. Benzing, R., Hopkins, V., *A catalogue of friction and wear devices*, ASLE May, 1966.

30. Bieber, H. E., Klaus, E. E., Tewksbury, E. J., *A study of tricresylphosphate as an additive for boundary lubrication*, ASLE Trans., **11**, 155–161 (1968).

31. Bisson, E. E., Johnson, R. L., Swikert, M. A., *Friction, wear and surface damage of metals as affected by solid films*, Proc. Conf. Lubr. and Wear, London, 384–391 (1957).

32. Biţă, O., Dincă, I., *Behaviour of phosphorus additives under heavy loads*, Rev. Roum. Sci. Techn.-Méc. App., **8**, 441–452 (1963).

33. Blok, H., *The flash temperature concept*, Wear, **6**, 483–494 (1963).

34. Blouet, J., *Tribométrie. Le frottement et l'usure*, Journées D'Etude, GAMI, 59–74 (1967).

35. Boes, D. J., Cunningham, J. S., Chasman, M. R., *The solid lubrication of ball bearing under high speed-high load condition from -225 to $+1000°F$*, Lubr. Eng., **27**, 150–159 (1971).

36. Boner, C. J., *Gear and Transmission Lubricants*, R. E. Krieger Publishing Co. Inc. Huntington N.Y., 1971.

37. Borsoff, V. N., Wagner, C. D., *Studies of formation and behavior of an extreme pressure film*, Lubr. Eng., **13**, 91 (1957).

38. Borsoff, V. N., *On the mechanism of gear lubrication*, J. Basic. Eng., March, 79–93 (1969).

39. Bowden, F. P., Tabor, D., *The Friction and Lubrication of Solids*, Oxford Univ. Press, London and New York, 1954/1964.

40. Braithwaite, E. R., *Solid Lubricants and Surfaces*, Pergamon Press, Oxford, 1964.

41. Brill, R., *X-ray and electron diffraction studies of lubricating surfaces*, Ann. N.Y. Acad. Sci., **53**, 824–835 (1951).

42. Brown, R. D., Burton, R. A., *Friction and adhesion of copper in vacuum*, J. Lubr. Technol., 425–431 (1967).

43. Brown, R. D., *Test Methods*, In F. F. Ling: *Boundary lubrication, an appraisal of world literature*, ASLE N.Y., 241–292 (1969).

44. Buckley, D. H., Johnson, R. L., *Lubrication in a vacuum*, Machine Design, **17**, 164–169 (1966).

45. Buckley, D. H., Robert, J. L., *Influence of crystal structure on friction characteristics of rare earth and related metals in vacuum to 10^{-10} mm of mercury*, ASLE Trans., **8**, *2*, 123–132 (1965).

46. Burton, R. A., Russell, J. A., *Forces and deformation of lead films in frictional processes*, Lubr. Eng., **21**, 227–233 (1965).

47. Cameron, A., *Die Theorie des Fressens von Zahnrädern*, Schmiertechnik, **14**, 136–139 (1967).

48. Chişu, A., Maros, T., *Studiul uzurii cu izotopi radioactivi*, Ed. tehnică, 1967.

49. Ciorăscu, F., Oncescu, M., *Determinarea şi măsurarea radiaţiilor nucleare*, Ed. Academiei, 1964.

50. Courtney-Pratt, J. S., Eisner, E., *The effect of tangential force on the contact of metallic bodies*, Proc. Roy. Soc., ser. A., **238**, 529–550 (1956).

51. Cucchiara, O., Goodman, P., *Kriptonates: a new technique for the detection of wear*, Materials Evaluation, **25**, 97–128 (1967).

52. Mc.Daniel, R. M., *The effects of reactor radiation on three high temperature solid-film lubricants*, Lubr. Eng., **21**, 463–473 (1965).

53. DAYSON, C., *Surface temperatures at unlubricated sliding contacts*, ASLE Trans., **10**, 169–174 (1967).
54. DEACON, R. F., GOODMAN, J. F., *Lubrication by lamellar solids*, Proc. Roy. Soc., ser. A., **243**, 464–482 (1958).
55. DEYBER, P., *Contribution à l'étude du coefficient de frottement et de la température en frottement mixte*, Thèse de docteur-ingénieur, Université de Lyon, 1970.
56. DINCĂ, I., BIȚĂ, O., *Influence de la nature du MoS_2 sur ses propriétés lubrifiantes dans le cas du contact linéaire et punctiforme*, Rev. Roum. Sci. Techn.-Méc. Appl., **14**, 1137–1155 (1969).
57. O'DONOGHUE, I. P., CAMERON, A., *Friction and temperature in rolling-sliding contacts*, ASLE Trans., **9**, 186–194 (1966).
58. DOWSON, D., *Transition to boundary lubrication from elasto-hydrodynamic lubrication, Boundary lubrication, an appraisal of world literature*, ASLE (1969).
59. FORBES, E. S., ALLUM, K. G., SILVER, H. B., *Load carrying properties of metal dialkyldithiophosphates. Application of electron probe microanalyses*, Proc. Inst. Mech. Eng., 1968–1969, Vol. 183, Pt. 335–49.
60. FUREY, M. J., *Metallic contact and friction between sliding surfaces*, ASLE Trans., **4**, 1–11 (1961).
61. FUREY, M. J., *Surface temperatures in sliding contact*, ASLE Trans., **7**, 133–146 (1964).
62. SAGEL, K., *Tabellen zur Röntgenemission und Absorbtionsanalyse*, Springer Verlag, Berlin–Göttingen–Heidelberg, 1959.
63. GÄNSHEIMER, J., *Influence of certain vapours and liquids on the frictional properties of molybdenum disulphide*, ASLE Trans., **10**, 390–399 (1967).
64. GÄNSHEIMER, J., *Neue Erkentnisse über die Wirkungsweise von Molybdändisulfid als Schmierstoff*, Schmiertechnik, **11**, *5*, 271–280 (1964).
65. GARKUNOV, D. N., KRAGELSKII, I. V., *On atomic capture from solids during friction*, Doklady Akad. Nauk., USSR, **113**, 326–327 (1957).
66. GODET, M., *Corrélations et divergences des méthodes d'essai du pouvoir lubrifiant des huiles*, Ing. Auto., **12**, 625–636 (1967).
67. GODET, M., FRÊNE, J., a.o. *Effets mécaniques introduits par la formation et la présence de films en surface*, Les mémoires techniques du CETIM, **11**, 8–14 (1972).
68. GODFREY, D., *Chemical changes in steel surfaces during extreme pressure lubrication*, ASLE Trans., **5**, 57–66 (1962).
69. GODFREY, D., *Boundary Lubrication*, Intern. Symp. Lubr. and Wear, Houston, Editor D. Muster, June 1963.
70. GODFREY, D., *The effect of high temperature on friction and wear*, Proc. Fifth World Pet. Congress, N.Y. section VI, Paper 23 (1959).
71. GODFREY, D., *The lubrication mechanism of tricresylphosphate on steel*, ASLE Trans., **8**, 1–11 (1965).
72. GOLDMAN, I. B., *Corrosive wear as a failure mode in lubricated gear contacts*, Wear, **14**, 431–444 (1969).
73. GREEN, E., *Review of surface texture measurement and the associated metrological problems*, Properties and metrology of surfaces, Paper 30, Inst. Mech. Eng., **182**, Pt. 316 (1967–1968).
74. GRÜNBERG, L., *The formation of hydrogen peroxide on fresh metal surfaces*, Proc. Phys. Soc. Sect., B, **66**, *3*, 153–161.
75. GÖTTNER, G. H., *Einführung in die Schmiertechnik*, Karl Marklein Verlag, Düsseldorf, 1961.
76. GÖTTNER, G. H., BARTZ, W., SCHULTZE, G. R., *Versuche über die Schmierwirkung von Molybdendisulfid*, Erdöl u. Kohle, Erdgas, Petrochem., **19**, 196–200 (1966).
77. GRATTAN, P. A., LANCASTER, J. K., *Abrasion by lamellar solid lubricants*, Wear, **10**, 453–468 (1967).
78. GREENHILL, E. B., *Adsorption of long-chain polar compounds from solution on metal surfaces*, Trans. Faraday Soc., **45**, 625 (1949).
79. GREENWOOD, J. A., WILLIAMSON, J. P. B., *Contact of nominally flat surfaces*, Proc. Roy. Soc., ser. A., **295**, 300–319 (1966).
80. HALTNER, A. J., *An evaluation of the role of vapor lubrication mechanisms in MoS_2*, Wear, **7**, 102–117 (1964).
81. HARRIS, C. L., READ, J. E., THOMSON, J. B., *Friction and performance of soft metal film lubricants in air and vacuum to 10^{-10} torr*, Lubr. Eng., **24**, 230–237 (1968).
82. HARRIS, C. L., WARWICK, M. G., *Lubrication of bearings and gears for operation in a space environment*, Proc. Inst. Mech. Eng., Vol. 183, Pt. 3 I, Paper (1968/1969).

83. HINSLEY, C. F., MALE, A. T., ROWE, G. W., *Frictional properties of metal oxides at high temperatures*, Wear, **11**, *3*, 233–238 (1968).

84. HOLM, R., *Electric Contacts, Theory and Application*, Springer Verlag, 1967.

85. HUENE-HOYNINGEN, A. (VON), *Aufbau und Wirkungsweise der wichtigsten Schmierstoffzusätze*, **3**, Tagung Schmierungstechnik, Wiesbaden, 1966. VDI-Berichte, VDI-Verlag, Düsseldorf, 111 (1966).

86. HUGEL, G., *Chemical nature of extreme pressure lubrication*, Lubr. Eng., **14**, 523–526 (1958).

87. HUGHES, J. R., RUSHTON, S. G., *Prüfung und Entwicklung von Einlaufölen für Hypoidgetriebe*, Erdöl und Kohle, **13**, 480–486 (1960).

88. HUGHES, J. R., *Shell 4-Ball E.P. Lubricant Tester*, Proc. Inst. Mech. Eng. Conf. on Lubrication and Wear, 1957.

89. HUGHES, J. R., *The Shell fourball E.P. lubricant tester. Method of use and precision in the determination of the E.P. properties of lubricants*, Proc. Gen. Discussion on Lubrication and Wear, Inst. Mech. Eng., London, 575–584 (1957).

90. ILIUC, I., *Aspecte ale frecării în regim limită studiate pe o cuplă cu trei contacte lineare*, St. cerc. mec. apl., **22**, 755–765 (1966).

91. ILIUC, I., *Importanța purității lubrifiantului în reproductibilitatea măsurătorilor coeficientului de frecare*, St. cerc. mec. apl., **27**, 835–839 (1968).

92. ILIUC, I., *Comportarea la frecarea de alunecare a unor straturi subțiri și influența mediului*, Doctor's thesis, Inst. Politehnic, București, 1969.

93. ILIUC, I., *Effect of some thin layers on the friction process*, Rev. Roum. Sci. Techn.-Méc. Appl., **15**, 235 (1970).

94. ILIUC, I., *Influence de certains procédés d'usinage sur le frottement et l'usure des surfaces métalliques*, Mécanique-Matériaux-Electricité (Paris), "L'usure", 34–38 (1972).

95. ILIUC, I., *Ein Beitrag zur Klassifikation und Analyse der Verfahren für Reibungs- und Verschleißuntersuchungen*, Schmiertechnik, **18**, 193–196 (1971).

96. ILIUC, I., *Der Verlauf von Reibungszahl, Übergangswiderstand und Verschleiß in einem Reibpaar Stift auf Scheibe*, Erdöl u. Kohle, **26**, 31–137 (1973).

97. JOHNSTON, R. F., MOORE, J. W., *The burnishing of molybdenum disulphide onto metal surfaces*, Wear, **7**, 496–512 (1964).

98. JOHNSON, V. R., VAUGHN, G. W., *Investigation of the mechanism of MoS_2 lubrication in vacuum*, J. Appl. Phys., **27**, 1173 (1956).

99. JONES, M. H., HOVELLS, R. I. L., PROBERT, S. D., *Solid in static contact*, Wear, **12**, 225–240 (1968).

100. JONES, J. A., MOORE, H. D., SCARLETT, N. A., *Grease lubrication of ball bearings in helium*, Proc. Inst. Mech. Eng., Vol. 183, Pt. 3 I Paper 1, 1968–1969.

101. KAHLES, I. F., FIELD, M., *Surface integrity. A new requirement for surfaces generated by material removal methods*, Properties and metrology of surfaces. Paper 4, Proc. Inst. Mech. Eng., Vol. 182, Pt. 3 K (1967–1968).

102. KERRIDGE, M., LANCASTER, J. K., *The stages in a process of severe metallic wear*, Proc. Roy. Soc., ser. A., **236**, 250–264 (1956).

103. KNAPPWOST, A., *Physikalisch-chemische Reaktionseffekte bei der Feststoffschmierung mit Molybdändisulfid und mit Graphit-Einlagerungsverbindungen*, Naturwissenschaften, **12**, 474 (1965).

104. KRUPPKE, E., *Gütewerte für Hypoidöle, gemessen in LKW-Hypoidhinterachsen*, Teil I, II und III. Schmiertechnik, **12**, 21–26; 84–94; 160–162 (1965).

105. KUBASCHEWSKI, O., HOPKINS, B. E., *Oxidation of Metals and Alloys*, Butterworths, London, 1962.

106. LANCASTER, J. K., *Anisotropy in the mechanical properties of lamellar solids and its effects on wear and transfer*, Wear, **9**, 169–188 (1966).

107. LANE, T. B., *The flash temperature parameter, a criterion for assessing E.P. performance on the four-ball machine*, J. Inst. Pet., **43**, 181–183 (1957).

108. LAURENSON, L., HOLLAND, L., BAKER, M. A., *Degradation of lubricating fluids in a vacuum*, Proc. Inst. Mech. Eng., **183**, Pt. 31, Paper 11, 1968–1969.

109. LEACH, E. F., KELLEY, B. W., *Temperature. The key to lubricant capacity*, ASLE Trans., **8**, 271–285 (1965).

110. LECHNER, G., *Die Bestimmung der Tragfähigkeit von Ölen als Grundlage für die Berechnung der Fressichercheit von Getrieben*, Mineralöltechnik, **7**, 1–19 (1962).

111. LECHNER, G., *Zahnradschmierung*, Maschinenmarkt, **74**, 24–31 (1968).

112. Levine, O., Zisman, W. A., *Physical properties of monolayers adsorbed at solid-air interface*, Part I, J. Phys. Chem., **61**, 1068 (1957).
113. Loeser, E. H., Twiss, S. O., *Analysis of films formed by radioactive E.P. additives*, Lubr. Eng., **14**, 343–349 (1958).
114. Magie, P. M., *A review of the properties and potentials of the new heavy metal derivative solid lubricants*, Lubr. Eng., **22**, 262–269 (1966).
115. Male, A. T., *Metal oxides for high temperature lubrication. Friction and Lubrication in Metal Processing*, ASME 200–208, (1966).
116. Manea, Gh., Boșianu, A., Chișu, A., Săveanu, L., *Organe de mașini*, vol. II, Ed. tehnică, 1958.
117. Manea, Gh., Mladinescu, T., Udrea, C., *Influența lubrificației și a vitezei periferice asupra capacității portante a roților cilindrice cu dantură dreaptă din poliamidă indigenă*, Mase plastice, octombrie, 1970.
118. Manton, S. M., O'Donoghue, J. P., *Gear lubricant testing machines — are they useful?* Tribology, **1**, 146–149 (1968).
119. Matveevski, R. M., *Cetyrokharikovaya mashina KT-2 dlya oprdelenya kriticheskikh temperatur pleonki masla na mealle*, Izd. VINITNF, **32**, 57–88 (1957).
120. Menther, J. W., Tabor, D., *Orientation of fatty acids and soap films on metal surfaces*, Proc. Roy. Soc. (London), Ser. A, **204**, 514 (1950).
121. Ming-Feng, I., *Pyrolysis of zinc dialchyl phosphorodithionate and boundary lubrication*, Wear, **3**, 309–311 (1960).
122. Ming-Feng, I., *A new approach in interpreting the four-ball wear results*, Wear, **5**, 275–288 (1962).
123. Milne, A. A., *Experiments on the friction and endurance of various surface treatments lubricated with molybden disulphide*, Wear, **1**, 92–103 (1957/1958).
124. Mitchell, E., Davies, C., *Molten salt bath nitriding*, Metal Treatment and Drop Forging, **31**, 3–49 (1964).
125. Moore, D. F., *A history of research on surface texture effects*, Wear, **13**, 381–412 (1969).
126. Mould, R. W., Silver, H. B., Syrett, R. T., *Investigations of the activity of cutting oil additives. I. Organosulphur-containing compounds*, Wear, **19**, 67–80 (1972).
127. Myers, N. O., *Characterization of surface roughness*, Wear, **5**, 182–189 (1962).
128. Nădășan, Șt., *Încercări și analize de metale*, Ed. tehnică, 171–218.
129. Nakajima, K., Kawamoto, J., *X-ray study of frictional wear in metals*, Wear, **11**, 21–28 (1968).
130. Nica, Al., *Theory and Practice of Lubrication Systems*, Scientific Publications Ltd., London, 1969.
131. Niemann, G., Retting, H., Lechner, G., *Zur Prüfung von Getriebeölen im Zahnrad-Verspannungsprüfstand. Stand der Erfahrungen*, Erdöl und Kohle, **19**, 472–480 (1959).
132. Niemann, G., Lechner, G., *Die Erwärmung der Zahnräder im Betrieb*, Schmiertechnik, **14**, 13–20 (1967).
133. Panish, W., *Spektrochemische Analyse mit Röntgenstrahlen*, Phillips Tech. Rdsch., **12**, 393 (1956).
134. Pavelescu, D., Iliuc, I., Barbul, S., *Étude d'une méthode radioactive pour l'estimation des qualités lubrifiantes des huiles à additif au point de vue de l'usure des surfaces*, Acta IMEKO 1961, 24, RU-152.
135. Pavelescu, D., Iliuc, I., *Apparatus for determining the antiwear properties of oils and materials*, Wear, **8**, 133–142 (1965).
136. Pavelescu, D., *Evoluția concepțiilor asupra naturii frecării și uzării solidului deformabil*, Bul. Științ. Inst. Construcții București, **14**, 315–347 (1971).
137. Pavelescu, D., *Concepții noi, calcul și aplicații în frecarea și uzarea solidelor deformabile*, Ed. Academiei, 1971.
138. Peterson, M. B., Johnson, R. L., *Factors influencing friction and wear with solid lubricants*, Lubr. Eng., **11**, 325–331 (1965).
139. Phillip, K. F., Pritchard, C., *Adhesion of metals in hot oxidizing gases*, Proc. Inst. Mech. Eng., Vol. 183, P 3 I, Paper 4.
140. Porgess, P. V. K., William, H., *Surface re-orientation, friction and wear in the unidirectional abrasion of graphite*, Proc. Phys. Sci., **76**, 513.
141. Potts, H. R., *Electron microprobe study of wear involving thin sections*, Wear, **3**, 228–235 (1966).

142. Quinn, T. F. J., *The dry wear of steel as revealed by electron microscopy and X-ray diffraction*, Proc. Inst. Mech. Eng., 182 Pt. 3 N, Paper 24 (1967/1968).
143. Quinn, T. F. J., *The Application of Modern Physical Techniques to Tribology*, Butterworths, London, 1971.
144. Quinn, T. F. J., *The effect of "hot-spot" temperatures on the unlubricated wear of steel*, ASLE Trans., **10**, 158 (1967).
145. Quinn. T. F. J., *Oxidational wear*, Wear, **18**, 413–419 (1971).
146. Rabinovicz, E., *Influence of surface energy on friction and wear phenomena*, J. Appl. Phys., **32**, 1440–1444 (1961).
147. Rabinowicz, E., *Practical uses of the surface energy criterion*, Wear, **7**, 9–22 (1964).
148. Rounds, F. G., *Effects of additives on the friction of steel on steel. I. Surface topography and film composition studies*, ASLE Trans., **7**, 11–23 (1964).
149. Rounds, F. G., *Some effects of non hydrocarbon base oil constituents on the friction and surface coating formation obtained with three additives*, ASLE Trans., **11**, 19–30 (1968).
150. Rowe, G. W., *Frictional behaviour of boron nitride and graphite*, Wear, **3**, 274–285 (1960).
151. Rowe, C. N., Dickert, I. I., *The relation of antiwear function to thermal stability and structure for metal 0, 0-Dialkyphosphorodithionates*, ASLE Trans., **10**, 85–90 (1967).
152. Russell, J. A., a.o., *Boundary lubrication behaviour of organic films at low temperatures*, ASLE Trans., **8**, 48 (1965).
153. Sakurai, T., Ikeda, S., Okabe, H., *A kinetic study on the reaction of labelled sulphur compounds with steel surfaces during boundary lubrication*, ASLE Trans., **8**, 39–47 (1965).
154. Sakurai, R., Sato, K., *Study of corrosivity and correlation between chemical reactivity and load-carrying capacity of oils containing extreme pressure agents*, ASLE Trans., **9**, 77–87 (1966).
155. Savage, R. H., *Graphite lubrication*, J. Appl. Phys., **19**, 1 (1948).
156. Sayles, F. S., *The four-ball E. P. Tester, an ASTM method of test*, NLGI Spokesman, August 1968, 162–167.
157. Smalheer, C. V., Mastin, T. W., *Lubricant additives and their action*, J. Inst. Pet., **42**, 337–347 (1956).
158. Solomon, G., *Grundlagen der Feststoffschmierung*, Mitteilung, Nr. 232, Zentrallaboratorium, TNO, Delft.
159. Steijn, R. P., *Friction and wear of single crystals*, in P. J. Bryant (Ed.), *Mechanisms of Solid Friction*, Elsevier, Amsterdam, 1964.
160. Steijn, R. P., *The sliding of surface of polytetrafluorethylene: an investigation with the electron microscope*, Wear, **12**, 193–212 (1968).
161. Stock, A. J., *Evolution of solid lubricant dispersion on a Four-Ball Tester*, Lubr. Eng., **22**, 146–152 (1966).
162. Studt, I., *Schmieröladditive für hohe Flächenpressung und ihre Wirkung*, VDI Berichte, 156 (1970).
163. Studt, P., *Einige Untersuchungen zur mechano-chemischen Reaktion organischer Chlorverbindungen sowie von Phosphorsäuretrio-kresylester mit Eisenoberflächen*, Erdöl und Kohle, **21**, 340–344 (1968).
164. Studt, P., *Zusammenhang zwischen Absorbierbarkeit und Wirksamkeit von Hochdruckzusätzen in Schmierölen*, Erdöl und Kohle, **21**, 784–785 (1968).
165. Summers-Smith, O., *A review of chemical effects*, Proc. Inst. Mech. Eng., Vol. 183, Pt 3 I, Paper 5 (1968/1969).
166. Tabor, D., *Solid friction and boundary lubrication*, Proc. Inst. Mech. Eng., London, vol. 182, Part 3 A, Paper 61, 1967–1968.
167. Tallian, T. E., Chiu, Y. P., Huttenlocher, D. F., Kamenshine, J. A., Sibley L. B., Sindlinger, N. E., *Lubricant films in rolling contact of rough surfaces*, ASLE Trans., **7**, 109–126 (1964).
168. Tenwick, N., Earles, W. E., *A simplified theory for the oxidative wear of steels*, Wear, **18**, 381–391 (1971).
169. Theyse, F. H., *Die Blitztemperaturhypothese nach Blok und ihre praktische Anwendung bei Zahnrädern*, Schmiertechnik, **14**, 22–28 (1967).
170. Thornton, H. R., Wolanski, Z. R., *Lubrication with electrodeposited films of Silver-Rhenium and Gold-Molybdenum*, Lubr. Eng., **23**, 271–272 (1967).

171. Tilenschi, S., *Chimie coloidală*, Ed. tehnică, Bucureşti, 1964.
172. Trapnell, M. W., *Chemisorption*, Butterworths, London, 1955.
173. Trillat, I. I., *La microsonde électronique et ses applications à l'étude et à l'analyse des matériaux*, Nucleus, **6**, (1966).
174. Vinogradov, G. V., Morozova, O. E., *A study of the wear of steel under heavy loads with lubricants containing sulphurbased additives*, Wear, **3**, 297—308 (1960).
175. Vinogradov, G. V., Korepova, I. V., *Smazocynoe deistvye nizkomolekulyarnkh uglerodov pri tyajelykh rejimakh trenya*, Teorya cmazornogo deistvya i novye materyaly, Izd. Nauka, Moskva, 1965.
176. Wagner, K., *Die Anwendung von Radionukliden für Verschleiß und Schmierungsuntersuchungen*, Isotopenpraxis, **4**, 85—94 (1968).
177. Wasem, R., Diergarten, H., *Wirkung unlegierter und legierter Mineralöle auf die Grenzfläche Werkstoff-Schmierstoff bei Rollreibung, insbesondere elektronenoptische Untersuchungen mit Zielpreparationen*, Teil I, Schmiertechnik, **9**, 69—78 (1962); Teil II, Schmiertechnik, **9**, 138—150 (1962); Teil III, Schmiertechnik, **9**, 200—206 (1962).
178. Waterhause, R. B., *The formation, structure and wear properties of certain non-metallic coating on metals*, Wear, **8**, 421—447 (1965).
179. Waterfall, F. D., *Reducing scuffing and wear of ferrous metals*, Engineering, **23**, 116—120 (1959).
180. Williamson, J. B. P., Hunt, R. T., *The microtopography of solid surfaces*, Burndy Research Rep., **59**, (1968).
181. Winer, W. O., *Molybdenum disulphide as a lubricant: a review of the fundamental knowledge*, Wear, **10**, 422—452 (1967).
182. Zisman, W. A., *Friction, durability and wettability properties of monomolecular films on solids*, Friction and Wear (Ed. R. Davies), Detroit 1957, Elsevier Publishing Company, Amsterdam, 1959.
183. Zlatin, N., Snider, E. R., *Critical aspects in the selection of cutting fluids for difficult-to-machine materials*, Lubr. Eng., **7**, 287—292 (1965).
184. * * * DIN 51354 1970, *Mechanische Prüfung von Getriebeölen in der F.Z.G.-Zahnrad-Verspannungs-Prüfmaschine.*
185. * * * DIN 513150 1971, *Bestimmung der Schweißkraft von flüssigen Schmierstoffen.*
186 Fed. Test Method, Std, nr. 791 a/6506 and 6507-T.
187. Fed. Test Method Std. 791 a/6501. 1, *Load-carrying capacity (SAE).*
188. Fed. Test Method Std. nr. 791 a/6505, *Load-carrying capacity of universal gear lubricants by the Timken machine.*
189. Fed. Test Method Std. nr. 719 a/6503, *Load-carrying capacity* (Mean Hertz Load).
190. IP/IAE Gear Rig Sub-Panel. *The four-ball machine; consideration of the repeatability and discrimination of different tests.*
191. The I.P. Gear Lubricant Rig Tests, Sub-Panel, *An investigation into the correlation between the IAE and FZG gear type lubricant tests machines*, J. Inst. Pet., **52**, 98—107 (1966).
192. *Les Traitements de Surface contre l'Usure*, Dunod, Paris, 1968.

Supplementary bibliography

193. Bos, A., *The temperature at the wear scars of the four-ball apparatus*, Wear, **31**, 17—27 (1975).
194. Peklenik, J., *New developments in surface characterization and measurement by means of random process analysis*, Proc. Inst. Mech. Eng., **182**, Pt. 3K, Paper 24 (1967—1968).
195. Arnell, R. D., Herod, A. P., Teer, D. G., *The effect of combined stresses on the transition from mild to severe wear*, Wear, **31**, 237—242 (1975).
196. Grevé, A., *Die wichtigsten Verschleissmethoden der Isotopentechnik*, Kerntechnik, **14**, 5, 204—209 (1972).
197. Beagley, T. M., *Laboratory assessment of a wear measurement technique using thin layer activation*, Wear, **32**, 117—121 (1975).

198. Czichos, H., *Failure criteria in thin film lubrication: the concept of a failure surface*, Tribology, **7**, 14–20 (1974).
199. James, D. H., Smart, R. F., Reynolds, J. A., *Surface treatments in engine component technology*, Wear, **34**, 373–382 (1975).
200. Lefevre, R., *Graissage et Tribotechnique*, Tome I, *Lubrifiants et processus de graissage*, Institut Français du Pétrole, Paris, p. 294.
201. Buckley, D. H., Pepper, S. V., *Elemental analysis of a friction and wear surface during sliding using Auger spectroscopy*, ASLE Trans., **15**, 252 (1972).
202. Bird, R. J., Galvin, G. D., *The application of photoelectron spectroscopy to the study of E.P. films on lubricated surfaces*, Wear, **37**, *143* (1976).
203. Cameron, A., Gordon, A. N., Symm, G. T., *Contact temperatures in rolling/sliding surfaces*, Proc. Roy. Soc., ser. A, 286, 45–61 (1965).
204. Fodor, J., Leib, I., Muk, G., *Measuring of the temperature maxima and distribution on the surface of dynamically heated moving systems using ^{85}Kr techniques*, Inst. Phys. Conf. Ser., No. 26, 1975, Chapter 7.

Subject index

See pg 20